Ann Fletchall / Chris Lukinbeal / Kevin McHugh
Place, Television, and the Real Orange County

AF719833

MEDIA GEOGRAPHY AT MAINZ

Edited by

Anton Escher
Chris Lukinbeal
Stefan Zimmermann

Volume 2

Ann Fletchall / Chris Lukinbeal / Kevin McHugh

Place, Television, and the Real Orange County

Franz Steiner Verlag

Cover illustration: Laguna Beach, California.
Photography by Ann Fletchall

Bibliographische Information der Deutschen Bibliothek
Die Deutsche Bibliothek verzeichnet diese Publikation in der Deutschen Nationalbibliographie; detaillierte bibliographische Daten sind im Internet über <http://dnb.ddb.de> abrufbar.

ISBN 978-3-515-10118-9

Jede Verwertung des Werkes außerhalb der Grenzen des Urheberrechtsgesetzes ist unzulässig und strafbar. Dies gilt insbesondere für Übersetzung, Nachdruck, Mikroverfilmung oder vergleichbare Verfahren sowie für die Speicherung in Datenverarbeitungsanlagen.
Koordination und Satz: Christina Kerz
© Franz Steiner Verlag, Stuttgart 2012
Gedruckt auf säurefreiem, alterungsbeständigem Papier.
Druck: Laupp und Göbel, Nehren
Printed in Germany

Geographers usually see themselves in a tradition of Alexander von Humboldt and we would like to highlight his position that everything in our world is interdependent. Deriving from that assumption geography has to be seen as an unlimited science, a discipline that knows its borders but has none.

Anton Escher, Chris Lukinbeal and Stefan Zimmermann,
Mainz, January 2008

CONTENTS

LIST OF FIGURES

CHAPTER 1

PLACE-MAKING AND THE MEDIA

Manhattan, to *Sex and the City* fans, is a collection of fashionable bars, restaurants, busy sidewalks, and iconic skyscrapers. Los Angeles, to viewers of *Beverly Hills 90210* or *The Hills,* is "a simple equation of sunshine, beaches, palm trees," trendy boutiques, and nightclubs (STENGER 2001, 63). To the millions who watched *Dallas,* it is to this day a city of oil money and big hair (CURRY 2004). Places created by television are more-than-real and larger than life. Although fragmented and exaggerated, the places television creates are more beautiful, more glamorous, and even more grotesque than residential experience, a vacation, or a business trip might lead one to believe. "It is not the allure of the real New York City that gives it such prominence," affirm SADLER and HASKINS (2005, 213), "but rather the appeal of the absolute fake version." Some places exist not as lived in, walked through, worked on, or "been there," but as they are seen on-screen. The evocative images that film and television produce create place; mediated place in simulated and stimulating forms is constitutive of contemporary experience.

Orange County, California, is a locale that is experienced in myriad ways. In our hypermediated culture, a neat and tidy distinction between direct and indirect (or mediated) experience cannot be sustained. Place-making and meanings are irretrievably bound up in media, including the twin towers of visual media: film and television. In exploring three popular television shows, we argue that the media, specifically television, plays a significant, yet overlooked, role in place-making.

Until recently, Orange County had limited exposure in film and television, with few movie roles and nary a *Sex and the City* to its credit. With the broadcast of three highly popular TV shows bearing its name, the County of Orange has been transformed from lackluster suburbia to televisual star. These shows—*The OC, Laguna Beach: The Real Orange County*, and *The Real Housewives of Orange County*—aided in transforming Orange County into the new and hip "OC," a glamorous spectacle, a more-than-real simulation. The transformation of Orange County into "the OC" illustrates how television moves beyond representing or simulating place by becoming an active participant in place-making.

This book fills a lacuna in geographical research: the role of television in place-making. In this introductory chapter we elaborate mediated place as process. We position the study within media studies in geography and articulate our perspective through an interrogation of key issues in place and representation. En route we illuminate the optical unconscious, animate place, implode the real/reel binary, tarry on the textual metaphor, and travel to televisual place as simulacra.

MEDIA GEOGRAPHIES: FAST FORWARD TO THE OPTICAL UNCONSCIOUS

Early forays in media geographies focused on mimetic and pedagogic qualities of film (MANVELL 1956; KNIGHT 1957), and pioneering work by Eugene WIRTH (1952) examined relationships between narrative, space, and place in film. It was not until the "cultural turn" in the late 1980s that geography began a serious engagement with visual media. Geographers in this period were initially attracted to visual media because this provided a way to examine non-material culture and representations. Both AITKEN and ZONN (1994) and NATTER and JONES (1993) make this argument in contending that geographers have traditionally emphasized material conditions over representations, set within the larger debate over "traditional" versus "new" cultural geography (COSGROVE 1993; DUNCAN 1993; JACKSON 1993; PRICE and LEWIS 1993a, 1993b). This debate focuses on whether "traditional" cultural geography, modeled after Carl Sauer and the Berkeley School, privileged material conditions and "fieldwork" to the extent that non-material cultural forms and representations were neglected.

Prior to the cultural turn, humanistic geography's engagement with media focused primarily on literature (MOORE and GOLLEDGE 1976; SALTER and LLOYD 1976; POCOCK 1981a, 1981b; MALLORY and SIMPSON-HOUSLEY 1987) and art (REES 1973; KING 1990; TUAN 1990), in part because film and television were viewed as forms of mere entertainment. By ignoring such a wealth of information, geography appeared to be an elitist pursuit wherein popular culture was not deemed worthy of inquiry (GOLD 1974; BURGESS and GOLD 1985).

Since these early forays into media, numerous books have engaged geographical implications of media (BURGESS and GOLD 1985; ZONN 1990; AITKEN and ZONN 1994; CLARKE 1997; SHIEL and FITZMAURICE 2001; CRESSWELL and DIXON 2002; JANCOVICH, FAIRE, and STUBBINGS 2003; LUKINBEAL and ZIMMERMANN 2008; ADAMS 2009). In these books, as well as in journal articles, cinema and space has garnered far more attention than television and place, which is surprising given the fact that in 1990 BURGESS outlined an agenda for media research in geography which identified the importance of television.

Our goal is to demonstrate ways in which television creates and perpetuates meaningful experiences vis-à-vis place. To do so we investigate how the techniques that form televisual landscapes structure narrative content and affect active engagement with place. These techniques, or narrative conventions, work to situate place within a landscape that promotes geographic realism, an effect that appropriates social-spatial meanings to a particular locale. Geographic realism works to authenticate place by grounding the fictional narrative to a "real historical place" (HIGSON 1984, 3). Geographic realism seeks to build and reinforce the social and spatial meanings of place. Geographic realism functions fully within the realm of optics, or as HIGSON (1987, 10) explains:

> Television constructs a sense of a public sphere of knowledge, action, events, and people, out there, separate from the viewer who observes from the private sphere of home. But at the same stroke, television constructs the public sphere *as a series of images*, in which the viewer *does* participate, precisely as observer (HIGSON 1987, 10).

Place-making requires more than mere observation and a voyeuristic engagement with the aesthetic production of televisual place. As BRUNO (2002, 16) notes, an optical model of filmic space reduces spectatorship to a fixed, disembodied gaze which is unfit "to account for the types of displacement that are represented, conveyed, and negotiated in the moving image." Optical models are important in understanding the power of the gaze as it naturalizes ways of seeing. A focus on optics, however, limits the inquiry to socially constructed meanings found within television. Film and television produce both a visual aesthetic and an "anti-aesthetic interruption" (DUBOW 2004, 277) that engages the optical unconscious. The optical unconscious "lies below the formal level at which the visual might begin" (DUBOW 2004, 266). In other words, we do not see vision but experience it through corporeal experience (DOEL and CLARKE 2007). The optical unconscious accentuates

> the conviction that the world is not a ready-made that can be counted on and reflected upon. It is an event, a happenstance, a taking place. What takes place, however, is not arbitrary, as eventfulness is increasingly *anticipated* by a plethora of more or less imperceptible 'performative infrastructures' (DOEL and CLARKE 2007, 898).

DUBOW (2004, 270) argues that the optical unconscious is "[s]tructured in mobility, it designates a space that eludes legibility, or rather it produces a space in which the visual struggles to 'take place.'" Spectators are not motionless subjects chained to identification with the camera's gaze; rather, they are engaged in a "kinetic affair" that exceeds the representation presented (BRUNO 1992, 115). This kinetic affair links the pleasures of spectatorship to the feeling of dynamic motion, which is generated by the camera's wandering eye. This pleasure differs from the idea of the immobile subject experiencing pleasure as a voyeur. We are not passive vessels experiencing voyeuristic pleasure; rather, we explore the presentation as tourist. Exploration is a kinetic moving gaze which transverses the televisual world as an extension of our world. This anti-optical engagement "is a haptical mobilization, an (e)motional journey that leads us to question our identity and its relation to values, experiences, and knowledge" (LUKINBEAL and ZIMMERMANN 2008, 21; cf. BRUNO 2002). The optical unconscious "provokes tension, contestation, and emotional responses because it does not separate subject (the viewer) from object (the content and form of the visual/audio)" (LUKINBEAL and ZIMMERMANN 2008, 21). Where the techniques of geographic realism work to construct a landscape in which places have meaning, the optical unconscious produces an emotional realism that breaks down the barriers between subject and object, allowing for place-making to occur. BRUNO (2007, 23–4) describes cinema as an "art of memory":

> [F]ilm itself draws memory maps. In its memory theater, the spectator-passenger, sent on an architectural journey, endlessly retraces the itineraries of a geographically localized discourse that sets memory in place and reads memories as places. As this architectural art of memory, filmic site-seeing … embodies a particular mobile art of mapping: an *emotion*al mapping.

Where place has been traditionally defined by its absolute location within a global graticule, or in relation to other places in a relative space, newer theorizations conceive place as an ongoing process, continually practiced and performed (MASSEY 2005). The traditional definition of place limits television to re-presenting place, while the latter challenges us to reconsider the role of visual media in the place-making process.

A PLACE FOR TELEVISION

The emergence and success of television as a mode of reception can be partially linked to the high rate of family formation following World War II in the United States. SPIGEL (1992) suggests that the growth of television in post-World War II was, in part, a cultural return to family values, neighborhood bonding, and community participation. Urban flight and rapid suburbanization in the 1950s spatially characterized this sociological change. Suburbanization and the advent of television forced a redistribution of consumption sites. SPIGEL (1992) links the rise of television to the values and ideology associated with suburban life. She argues that "suburbia emerges as a conformist-oriented society where belonging to the neighborhood network was just as important as the return to family life" (SPIGEL 1992, 86). Television aided in the re-visioning of this morality by providing a means through which people could safely negotiate spatial boundaries between the public and private sphere. Television provided a form of access to the sphere of the public within the private home. Television promised "familial bliss" which was "wholesome" in that it was white, middle class, and heterosexual (SPIGEL 1992, 192). Television purified public space by standardizing and naturalizing the "American experience." In short, television provided unity through cultural hegemony.

Watching TV remains one of America's favorite pastimes, and television is certainly one of the most ubiquitous of mediums. In addition to the television screens which populate airports, sports bars, and health clubs, the average American home has more TV sets than people. During the 2006–2007 television season, American households watched, on average, eight hours of TV per day (Nielsen Media Research 2007). TV is, undeniably, a significant part of our lives.

Cinema has received much more attention in scholarly circles than television. This is due, in part, to the smaller and squarer TV screen and the fact that TV shows are shorter and interrupted by commercials, often failing to command the full attention of viewers (HIGSON 1987). While movies are better conveyors of sense of place, owing to a superior aspect ratio for depicting landscapes, television's episodic nature strengthens its place-making power. As opposed to a singular movie seen once or perhaps a few times at most, viewers make weekly visits to the world of the TV show. If the show proves successful, these visits may extend over a period of years. The structure of a TV series allows *more* time for place to be created and for viewers to develop attachments to characters and become emotionally involved in their lives. Televisual place displays an aura of familiarity. The act of viewing can become a highly anticipated event, with escapes to the televisual world being an integral part of one's weekly routine.

PLACE AND TELEVISION

We move away from fixed and concretized notions in favor of a more animated view of place as gatherings, encounters, events, entanglements (THRIFT 1999; MASSEY 2005). Place is constantly produced and reproduced in a world in movement, open to change, mediation and contestation (CRESSWELL 2004). Place is a calling forth, a putting to task through practice and performance, akin to what Tim INGOLD (1993) calls a taskscape. An animated conceptualization does not imply that place is without structures, but rather that structures are emergent in ongoing practices and performances which make place meaningful. Place is a gathering—a gathering of things, thoughts, memories and emotions in particular configurations (ESCOBAR 2001).

In his treatise on television, ADAMS (1992, 118) conceives of place as a "system in which symbolic interaction among persons occurs (a social context), and a nucleus around which ideas, values, and shared experiences are constructed (a center of meaning)." ADAMS (1992) argues that television provides a social context that unites a great number of people around a common experience or event such as the Superbowl or the World Cup. Television also produces cultural symbols and meanings through the introduction and popularization of, say, Lady Gaga. In ADAMS' (1992, 119) words, "insofar as people experientially inhabit it and relate to other persons through it or within it," then television is a place, or more precisely a "gathering place" (ADAMS 1992, 117). Televisual gathering places are not universal, but rather are contested domains in which meaning, power and knowledge are naturalized and contested. MASSEY (1994, 154) similarly suggests that place is the particular intersection of networks of social relations, movements and communications. For both MASSEY and ADAMS, places are not spatially defined areas with distinct boundaries, but are "articulated moments in networks of social relations" (MASSEY 1994, 154).

Meeting or gathering place has many forms; it cannot be restricted to the material embodiment of social processes in a particular location. Place is an expression of emotional resonance, a product of our imagination, memories and image-events from television and cinema. André MALRAUX (1967) describes place as a *musée imaginaire*:

> We … carry around with us a *musée imaginaire* in our minds, drawn from experience (often touristic) of other places, and knowledge culled from films, television, exhibitions, travel brochures, popular magazines, etc. It is inevitable … that all of these get run together (HARVEY 1989, 87; cf. JENCKS 1984).

According to the art historian Denis HOLLIER (1998, 64), MALRAUX's conception is "a museum conceived in terms of cinema."

Televisual places are becomings threaded with feelings and emotion. Whereas space is often conceived in objective fashion, place speaks to meaningful experience in the world. As TUAN (1977, 4) argues, "places are centers of felt value." BRUNO (2007, 38) takes TUAN's argument one step further in positing that cinema and television, through moving images, are an "intimate exploration – a screen of personal and social, private and public narratives" (BRUNO 2007, 38). It is thus a

> map of intersubjective views. A haptic architexture. A topophilic affair. A place for the love of place. A site of close picturing for undistanced *emotion*. A museum of emotional pictures (BRUNO 2007, 38).

Media are an indispensable component of place, contributing to how we experience worlds, not as universal, but differentiated by social, cultural and subjective processes that make place meaningful (RELPH 1976). In humanist terms, the essence of place cannot be defined by absolute location or functions, but rather through human existence and immersion (RELPH 1976). Places involve a "concentration of our intentions, our attitudes, purposes, and experience"; they are "the point of departure from which we orient ourselves in the world" (RELPH 1976, 43). Place is

> a way of seeing, knowing, and understanding the world. When we look at the world ... we see attachments and connections between people and place. We see worlds of meaning and experience (CRESSWELL 2004, 11).

In television and cinema,

> places are used as wax, they bear the layers of a writing that can be effaced and yet written over again, a constant redrafting. Places are the site of a mnemonic palimpsest ... places live in memory and revive in the moving image ... Mechanically made in the image of wax simulacra, the projected strip of celluloid is the modern wax tablet. Not only the form but the écriture is reinvented in film's own spatial writing, décor, and palimpsestic architectonics, as well in the spectatorial promenade. The loci of the art of memory bear the peculiar wax texture of a filmic "set" – a site of constant redrawing, a place where many stories both take place and take the place of memory (BRUNO 2007, 21–22).

IS TELEVISUAL PLACE A REPRESENTATION?

In geography there is a normative belief that television and cinema are representations of some ontologically stable reality awaiting our exploration. As AITKEN and DIXON (2006, 327) explain, "we can no longer talk of film representing, or mimicking, reality because we can no longer assume that there is a single reality waiting out there to be filmed." They go on to argue that while the "camera records mass and motion ... the 'nature' of the objects that appear on screen is firmly located in the social realm wherein meaning is ascribed to them."

We can ascribe two competing theories about the role of place within television and cinema. On the one hand, the normative belief ascribes to a logic wherein television and cinema are representations and thus can be assessed by their accuracy or success in providing "unmediated" access to the real world (MANVELL, 1956; KNIGHT, 1957; HARVEY 1989). Exemplary of this approach is HARVEY's (1989) use of film in illustrating the condition of postmodernity. He sees certain films as "brilliant portrayals ... of the conditions of postmodernity, and in particular the conflictual and confusing experiences of space and time" (HARVEY 1989, 322). HARVEY observes film merely as an apparatus that projects pictures in motion which reduces "the complex stories of daily life to a sequence of images upon a depthless screen" (HARVEY 1989, 322). Because of this, no film "has the power to overturn established ways of seeing or transcend the conflictual conditions of the moment" (HARVEY 1989, 322). HARVEY perceives film as capable of only mirroring the conditions of lived experience (AITKEN and ZONN 1994). As he puts it, "in the final analysis" film is "a spectacle projected within an enclosed space on a depthless screen" (HARVEY 1989, 308). For HARVEY, material conditions should always be privileged over representations. Here, representations of place are tied to the re-presentation of absolute locales where mediated place is no more than a vicarious or secondhand experience (LUKINBEAL and ZIMMERMANN 2008).

A second view holds that place is a fusion of the actual and virtual (or mediated), as the actual and virtual are two interrelated dimensions of the real (MARTIN, 2000). This perspective finds expression in varying ways in cultural and geographical thought, including SOJA's (1996, 11) influential work articulating a geographical imaginary as a "thirdspace" that is "simultaneously real and imagined." Following the logic of SOJA, HANNA (1996, 638) argues that place is "always-already a representation." His examination of Roslyn, Washington, – which played "fictional" Cicely, Alaska, in television's *Northern Exposure* – demonstrates how a particular locale is a fusion of the real and imagined. In *Northern Exposure*, Roslyn constitutes the real, while Cicely is the imaginary; the locale is mutually constituted by both, as well as by the cultural history of the town. HANNA (1996) argues that this place is paradoxical in that Roslyn and Cicely can never wholly be constituted as either real or imagined because all places have a representational legacy that extends beyond practices that produce their current configuration. Prior to its notoriety as Cicely, Roslyn's built form and identity was tied to its history as a company mining town, its revitalization as a quaint historic site, and subsequent evolution to vacation spot for Seattle-ites. Roslyn's identity is bound to these legacies. The creation of televisual Cicely, from Roslyn, adds another layer to the representational mélange that is this place. In HANNA's (1996, 642) words: "Roslyn, as a place, is just as much a representation as Cicely is material."

HANNA's (1996) essay on Cicely/Roslyn seeks to engage place as mutually constituted by its representational legacy and its material landscape. Its role as Cicely added a layer to Roslyn's current form, including *Northern Exposure*-related gift shops and remnants from the set, but HANNA (1996) does not consider the televisual place that *Northern Exposure* has created. We see that Cicely is a part of Roslyn, but what of the meaningfulness of Cicely's Alaska?

While HANNA offers one way to address how place-making incorporates the real and imagined, it remains cast within the real/reel binary and thus elaborates only how a morphological approach can be applied in understanding televisual place as a representation. A morphological approach requires the stripping away of layers to uncover the various meanings inscribed in a locale. As AITKEN and DIXON (2006, 329) note:

> This is not necessarily articulating landscape [or location] as a passive stage upon which culture struts its stuff. Rather, the morphology of landscape is seemingly depicted accurately in the ways it actively connects with culture. In this interpretative gambit, culture is a factor and landscape is a medium.

As such, belief in the reel/real binary is reinforced as cultural meaning is naturalized and inscripted onto an ontologically stable locale, rather than seeking to expound upon the meaningful processes in place-making. Furthermore, POORE and CHRISMAN (2006, 513) argue that the

> landscape-as-layers metaphor demonstrates how a discursive practice, which originated as a leap in scientific logic, can interact through time with the agencies of people, institutions and technologies to enforce certain ways of seeing and talking about the landscape.

These ways of seeing reinforce the idea that meaning is inscribed onto a pre-given landscape or location. COSGROVE (1989) similarly deploys the landscape-as-layer metaphor, reinforcing the morphological approach where different layers embody different symbols and meaning. The problem with the layered metaphor is that it "carries the implication that the way to uncover the most basic level of human beings' practical involvement with their environs is by stripping these layers away" (INGOLD 1993, 171). Rather than stripping away layers, or decoding sequential deposits of meaning, cinematic and televisual landscapes and places are never complete; they are performative becomings.

TUAN (1991, 690–691), too, recognizes the influence of evocative literature upon the "real world" but maintains the binary distinction between real and imagined places:

> 221B Baker Street [of Sherlock Holmes fame] is more vividly present to some Londoners than are the apartment homes of their maiden aunts, and more real by far to tourists than are the hotels they temporarily occupy … [London] can seem unreal precisely because it is so thoroughly transformed by the literary imagination … [I]t goes without saying that a literary person's sense of reality is more thoroughly penetrated by what he or she has studied and absorbed.

The same might be said of the reality of an earnest television watcher; rather than by words from a novel, it is a reality permeated with images from the screen. Take for example the Manhattan Television Tour, which TORCHIN (2002) presents as an experience which thrives on blurring the boundary between the televisual and "real" worlds. On the tour, guests are shuttled around the city to glimpse what would otherwise be ordinary New York buildings, but for their use in establishing exterior shots of popular TV shows (stops include the *Friends* apartment building and *The Cosby Show* brownstone). Drawing from ROJEK (1997), TORCHIN (2002, 251) applies the concept of "dragging" to the tour: signs, symbols, and images in movies and TV "are dragged onto the physical landscape and the physical landscape is then reinterpreted in terms of [these] cinematic events." In other words, place unfolds as an amalgam, simultaneously real and mediated. Place emerges, not only by our presence in it, but with fragments and figments of our *musée imaginaire* and the thrill of association with the more exciting, simulated world of TV. Through all of these, reality is enhanced, and "the everyday world of New York City is reconfigured as interesting" (TORCHIN 2002, 248).

More recently, media geographies have turned to a radically different understanding of the ontology of film and television. CRESSWELL and DIXON (2002, 3–4) argue that visual media is not simply "mere images of unmediated expressions of the mind, but rather the temporary embodiment of social processes that continually construct and deconstruct the world as we know it." They go on to argue that the dominance of the textual metaphor in media geography reinforces the reel/real binary through its focus on text and context, implying a fixed positionality of the observer as she relates to a film or television show. The reel/real restricts inquiry of the spatiality of media. LUKINBEAL and ZIMMERMANN (2006, 322) challenge this normative belief in stating:

> We eschew the representational determinism that film geography is synonymous with *re*-presentations of some ontological stable "authentic" reality. Drawing from CRANG et al. (1999, 2), we posit that film geography always exceeds the cinematic technologies which produce representations because film is "constituted by the social relations, discourses and sites in which these technologies are embedded." The technologies that capture, encode and represent the world are always embedded in social and cultural practices that are temporally and spatially specific. Representations are not the polar opposite of reality especially when it comes to film and cinema. Cinematic images are always socialized just as technologies are always socialized.

The textual metaphor is a hermeneutical method that assumes that cinema and television are authored and can be read like texts. While humanistic inquiries into the textual metaphor (cf. LEWIS 1979; MEINIG 1979) focused on "natural" or an "automatic" reading of the landscape, new cultural geographers have problematized the meanings associated with "reading," "writing" and "text" (DUNCAN and DUNCAN 1988; DUNCAN 1990; BARNES and DUNCAN 1992; COSGROVE 1993; DUNCAN and LEY 1993). The textual metaphor presupposes that there is no single "author," for it is through the act of "reading" that production and reproduction occurs. While landscapes, films, and television shows are written and read by experts and ordinary people alike, they constitute and reflect power relations that exist in culture and society (DUNCAN and DUNCAN 1988). The textual metaphor does not offer a meta-narrative which can read landscapes, but rather suggests an intertextual world of cultural signifying practices whereby a text is understood, interpreted and constituted by relations with other cultural texts. Landscape, film, and television are intertextually related to other systems of cultural production and reproduction (cf. COSGROVE 1984, 1987, 1993; DANIELS and COSGROVE 1988; DANIELS 1989).

When applied to film analysis, the textual metaphor often has the effect of limiting spatial inquiry to internal textual properties. According to HAY (1997, 214), cultural criticism through textual analysis focuses on the "internal properties of texts or upon the relations among film texts" which suggests a "kind of placelessness of an absolutely separate place of the literary/filmic in social relations and history." Textual analysis, according to HAY (1997, 214), assumes that the reading of the text is done from "no place in particular" and that critical theory applied to a text is done from "anywhere." From this vantage point, the textual metaphor assumes a "counter-public sphere" from which we might "intervene in social relations without addressing the locational politics of such a sphere" (HAY 1997, 214). While HAY does not take into account that geographers have addressed the cultural politics of authorship (DUNCAN and LEY 1993; ROSE 1993), his argument does point to a gap in our understanding of the spatial politics of the textual metaphor.

By reducing film and television to text, it becomes an instrument and gives research the status of a reading. This not only fetishizes space by making the researcher's mental space envelop social and physical space (LEFEBVRE 1991), but it also sidesteps the history and social practices associated with the film production industry. The textual metaphor is a powerful and appropriate means of engaging spatial politics of the gaze and optical power inherent in film and television. But, we must realize that the textual metaphor has limits; there are extra-textual cultural processes that cannot be captured adequately via this metaphor, most notably the optical unconscious. The textual metaphor never truly offers what SMITH and KATZ (1993) call an *effective* and *coherent* spatialized politics because, according to HAY (1997), the locational politics of what is cinematic or televisual remains fixated on the narrative space within the text. Just as spatial metaphors work to hide cultural processes related to historical materialism (SMITH and KATZ 1993), the textual metaphor applied to film and television overlooks intertextual and extra-textual processes. It is our contention that cinema and television always exceed the space defined by the textual metaphor. To focus primarily on text in analyzing visual media runs the risk of losing touch with how television and cinema affect and engage everyday life.

TELEVISUAL PLACE AS SIMULACRAL

In his 1935 essay "The Work of Art in the Age of Mechanical Reproduction," Walter BENJAMIN addresses the profound effects of film on art and society, astutely recognizing early on that film would reconfigure traditional modes of representation. If art is a representation of reality, film creates a much different representation than a stage play or a painting: The movie camera penetrates reality so completely that it presents "an aspect of reality which is free of all equipment," says BENJAMIN (1968, 234), "and that is what one is entitled to ask from a work of art." Furthermore, as articulated by CLARKE (1997, 9), the audience sees from the camera's point of view such that film is not merely a "*representation* of space" but "*(re)produces* a virtual space" and, in effect, achieves a "re-framing of the world" (emphasis in original). The technology of film allowed us to see the world in novel ways, such as in close-up and in slow motion. "Cinema is at once a form of perception and a material perceived, a new way of encountering reality and a part of reality thereby perceived for the first time" (SHAVIRO 1993, cited in CLARKE 1997, 2). Indeed, from the comfort of home or as one in the crowd at a movie theatre, BENJAMIN (1968, 236) was prescient in arguing that film would change our experience of the world:

> Our taverns and our metropolitan streets, our offices and furnished rooms, our railroad stations and our factories appeared to have us locked up hopelessly. Then came the film and burst this prison-world asunder … so that now, in the midst of its far-flung ruins and debris, we calmly and adventurously go traveling.

The subsequent development of montage furthered the manipulation of space and time, so that film was freed from its role as a "referential medium, bound to the Real, to become a *simulacral* medium, free to fabricate a reality-effect" (DOEL 2008, 96). Montage now presents a "reality effect" such that "once assembled, a film does not re-present a world that preexisted it" (DOEL and CLARKE 2007, 897). Film and television create their own realities that provide an experience in and of themselves (AITKEN and ZONN 1994; CLARKE 1997; DOEL and CLARKE 2007).

BAUDRILLARD's postmodern theory jettisoned the belief that there is "an existing, knowable world," a truth to be revealed behind the media's glossy images (BEST and KELLNER 1997, 101). As a radical reconfiguration of Marxist theory, BAUDRILLARD saw that beyond both use value and exchange value, the commodity is purchased for its sign value, not for what it is, but for what it represents. Clothes and cars, rather than body coverings and transportation, have become success and glamour. What we see on television is what we strive to be, and our actions are conditioned by the movies. BAUDRILLARD claims that media images have so thoroughly permeated experience that we have lost any referent to the real; the image is reality (BEST and KELLNER 1997).

In webs of virtuality, we are left with nothing more and nothing less than a system of simulacra: "realer-than-real" constructions, depthless artifices, copies without an original, a "hyperreality." For BAUDRILLARD, the mass media have disturbed the ontology of the entire social world (BEST and KELLNER 1997). Unlike Adorno, BAUDRILLARD does not view the masses as defeated or duped, but as bored, over-stimulated, and unquestioning, and it is by ignoring the media's message that we, in effect, resist it. Caught as we are in worlds of simulacra, we are plunged "into a state of stupor": "a radical uncertainty as to our own desire, our own choice, our own opinion, our own will" (BAUDRILLARD 1988a, 209). We suffer from "a state of suspense and of definitive uncertainty about reality" (BAUDRILLARD 1988a, 210). All we can do is "absorb it without reply"; this is our silent act of defiance (POSTER 1988).

Worlds created by film press into everyday life. BAUDRILLARD (1988b, 55–56) explains:

> In fact, the cinema here is not where you think it is. It is certainly not to be found in the studios the tourist crowds flock to … Where is the cinema? It is all around you outside, all over the city, that marvelous, continuous performance of films and scenarios … The American city seems to have stepped right out of the movies … [cinema] invests the streets and the entire town with a mythical atmosphere.

In BAUDRILLARD's (1988b, 104) words, "The cinema and TV are America's reality!" In light of this revelation, LUKINBEAL and ZIMMERMANN (2006, 11) raise the question: "What if the copy is better than the original? What if the original never existed but is a myth?" In his examination of LA's Hollywood redevelopment project, STENGER (2001) demonstrates that there can be serious material repercussions when the simulated world of film is better-than-real. "Hollywood-the-place" is being made over so as not to disappoint tourists expecting to see "Hollywood-the-cultural-myth" (STENGER 2001, 71). The redevelopment of Hollywood Boulevard is well on its way to becoming a simulacrum of a simulacrum. Television and cinema are simulacral, which shifts the mode of inquiry away from the problematic reel/real binary to the openness afforded by mediated place. We are not passive viewers or voyeurs of movement-images; we are active itinerates participating in, and creating, meaningful experiences and connections vis-à-vis place.

PREVIEW OF COMING ATTRACTIONS

Our exploration of Orange County and the ways in which television and the media are intimately involved in place-making begins in chapter two. In this chapter, we trace the mediated and geographical imaginary of Orange County prior to its television popularity. This is followed in chapter three by an examination of the popular culture transformation of the County of Orange into the "OC" as a result of the shows *The OC, Laguna Beach: The Real Orange County*, and *The Real Housewives of Orange County*. These chapters provide background needed to position Orange County as a mediated place, highlighting ongoing processes of identity formation in mediated place-making.

In chapter four we address landscape and place-making, accentuating production techniques in television relating to geographic realism. In chapter five we reveal how emotional realism works to produce an affective response in televisual place-making. Chapter six draws on survey data to explore ways in which audiences experience "reality" TV shows. Here we highlight the tension between television as a re-presentation of Orange County itself and as an affective agent engendering emotional attachment to something called the "OC." Chapter seven delves into the tension between the optical reel/real place and the anti-optical emotional affect via a touristic experience by the lead author (Ann) in search of "the OC." We conclude in chapter eight with a discussion of the paradox of two places in one location: the County of Orange and the "OC" as mutually inclusive and exclusive.

CHAPTER 2

INTRODUCING ORANGE COUNTY

Orange County, California, is home to Disneyland and Little Saigon, the planned community and Surf City, U.S.A., modern conservatism and hardcore punk, the original Leisure World and the skateboarding industry, John Wayne, and Richard Nixon, is Orange County, California. Orange County is not a place whose essence is wholly captured with the adjectives often used to describe it: suburban, conservative, and wealthy. Beyond wealthy, its boundaries contain one of most expensive housing markets in the country – Newport Beach and its spectacular oceanfront monuments to conspicuous consumption (Coldwell Banker 2011). Mere miles inland, however, lies Santa Ana, a community recently ranked "the nation's toughest place to survive" due to poor living conditions and high rents experienced by its Mexican immigrant community (Anton and Mena 2004). Orange County is remarkably diverse, yet in stripes of this noteworthy stretch of land where such descriptors do ring true, they fail to do it justice. It is, for example, a gross simplification to label Orange County as merely suburban, when its shopping malls and planned communities are national models for development and its sprawl exemplifies a new urban form. Orange County doesn't just possess right-wing tendencies; its brand of politics spawned the contemporary conservative movement (McGirr 2001). Orange County is by no means a uniform and straightforward entity. It is a surprising place of multiple, even contradictory, images and identities, making it one of the most fascinating counties few may have paused to consider.

Place identity can be described as the image a place projects and is known by. Place identity might be cultivated by area boosters, adhered to by residents, or felt by tourists, but for those of us who only know a place from afar, its identity is a product of its mediations. This chapter paints a picture of mediated Orange County, gleaned from academic studies of its particular growth and development, fashioned from novels and films for which it serves as a setting, and through the cultural icons it has produced. We sketch several identities on this journey through Orange County – the American Dream, the anti-LA, and the Tragic Kingdom – which only together can begin to capture this sun-kissed, yet at times bleak, paradoxical land.

THE AMERICAN DREAM (AND ITS PROTECTION)

Orange County is a collection of thirty-four municipalities within the 948 square mile rectangle of land situated between the Pacific Ocean and the Santa Ana Mountains to the east, and between Los Angeles and San Diego Counties to its north and south (Figure 1). It is the second most populous County in California (U.S. Census 2000); although its southeastern third is protected by Cleveland National Forest, most of its residents reside farther north and nearer the coast (Figure 2). The area was colonized by the Spanish in the late eighteenth century, upon which enormous land grants were awarded to the king's most favored settlers. The land became Mexico's after winning independence from Spain in 1821 and then U.S. territory in 1848 following the Mexican-American War. California became a state in 1850. Under American legislation, southern California's great cattle ranches, the legacy of the Spanish land grants, were broken up and given over to agricultural pursuits, namely orange production (McGirr 2001). During the 1880s, residents of what was then southeastern Los Angeles County, fed up with contributing their tax dollars to fund the urban ills festering to the north, bribed LA and state officials to pass a bill allowing them to secede from Los Angeles County. On March 11, 1889, these "pioneers" voted to form Orange County (Till 1993, 720).

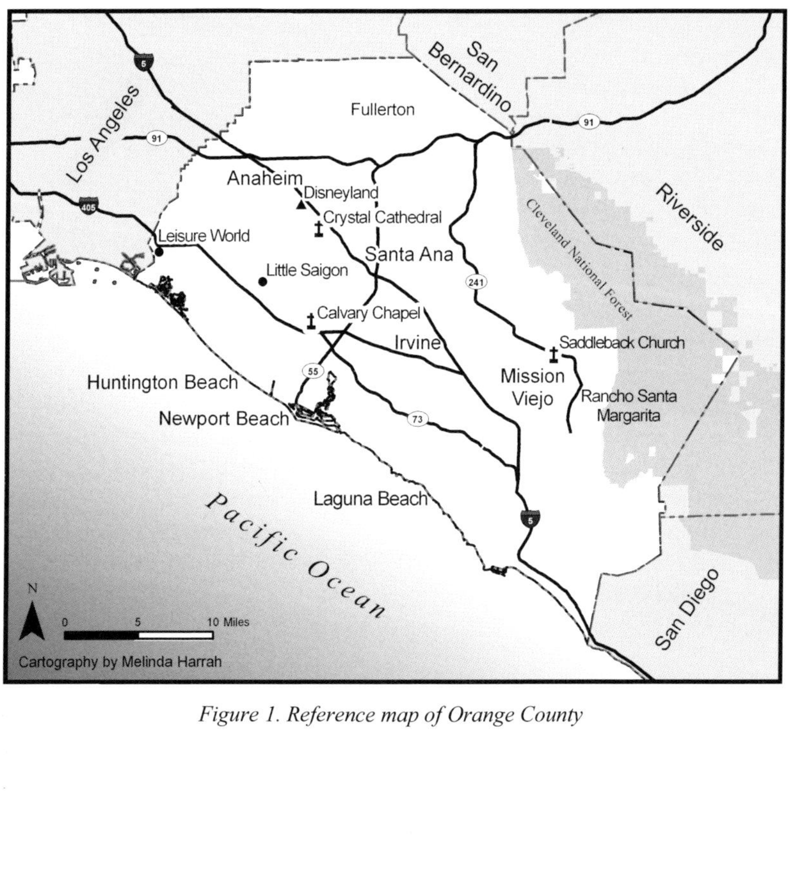

Figure 1. Reference map of Orange County

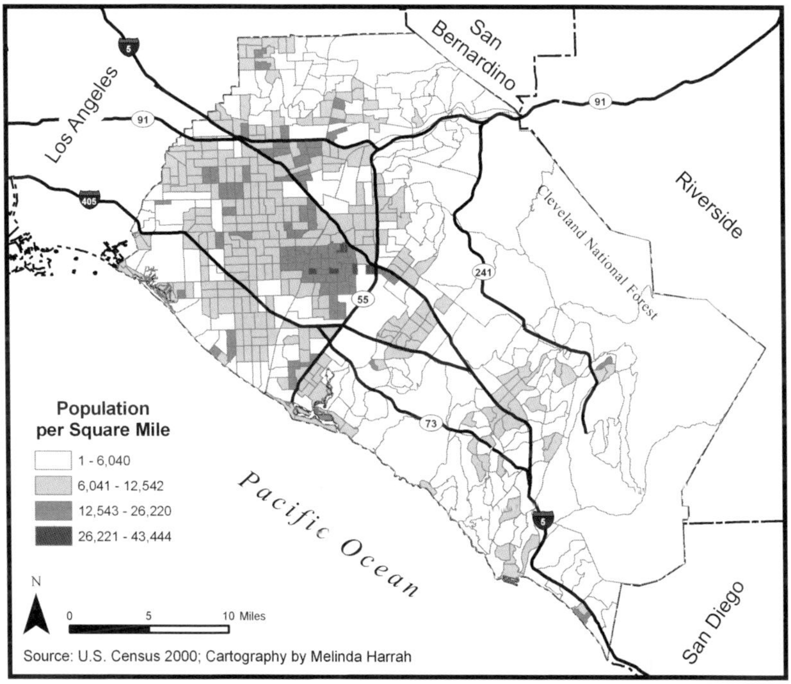

Figure 2. Orange County's population density

Orange County remained rural until the postwar period, upon which it was called to the front lines of the Cold War. Southern California became a major beneficiary of government defense spending; although LA was at the center of this boom, much business spilled over into northern Orange County, where there was still plenty of room to expand the military-industrial complex (SCOTT 1988; MCGIRR 2001). Migrants rushed into the area to fill these jobs and to provide support services for the burgeoning industry.

Orange County thus exploded. Its population grew 225% between 1950 and 1960 (from 216,000 to 704,000) and 101% between 1960 and 1970 (to 1.4 million). By the mid-1960s, the northern half of the county had developed into what SCOTT (1988) has famously titled "The Orange County High-Technology Complex," its centers congealing in Anaheim and Fullerton.

After such explosive growth, northern Orange County was no longer a quiet agricultural "backwater," but something of an urbanized region (SCOTT 1988, 166). It became a top destination for migrants from Middle America, and those who moved there realized some economic success and could afford a nice new home – this was, after all, the American Dream. Walt Disney, after the same dream-seekers, built his theme park in Anaheim in 1955, deducing a locational advantage in the heart of then-novel suburbia (MCGIRR 2001).

Not wanting to stymie development, the form of growth in northern Orange County was not unlike that of LA's Inland Empire. It was unplanned and unfettered, with little regard for public space or traffic abatement (MCGIRR 2001). Meanwhile, much of southern Orange County remained farm and ranchland. With a mission to avoid the sprawl, congestion, and general "LA-ness" which had befallen the northern portion of the county, development in the south in the 1970s and 80s was of a different character (TILL 1993). Entire cities were planned and developed by private corporations in accordance with a utopian master plan. The City of Irvine, for example, was built by the Irvine Company on land that was formerly the family ranch and designed to appeal to a homogenous group of upper-middle-class home-seekers. Southern Californians had one more shot at the American Dream. Andrea SEIGEL (2004, 10–13, 18) in her novel *Like the Red Panda,* offers an evocative description of Irvine's ubiquitous planning, as seen through the eyes of a disaffected high school girl (Figure 3):

> We live in a village of what is technically a city (I guess) called Irvine. Sometime around the sixties the Irvine family must have been taking their horses out for a spin, and one of them thought aloud, "A lot of people have cities named after them, but how many people have cities that turned out exactly how they imagined?" So I'm assuming they pulled out a piece of graph paper back at the ranch and drew up a dream plan, and they got really excited by the idea of seeing the future and the part they'd play in making it true. The Irvines plotted years ahead so that even Southern Irvine, the later part of the city to be developed, was already a gleam in North Irvine's eye by the time the original houses were built there.
>
> Being older, Deerfield (the extreme back end of North Irvine) is quieter than its newer counterpart and earthier too, I mean, as earthy as a place can get where the homes will still run you a half million … Deerfield even has a 7-Eleven, which the South wouldn't be caught dead with. The existence of a 7-Eleven says that someone somewhere got lazy.
>
> And that's the thing about Woodbridge, the village where I live now. When I hear people talk about upper-middle-class America, they rattle on about the mediocrity of the homes and the scenery and the lives contained within them. But Woodbridge was created with what I can only identify as intense passion. The Irvine family took stock of Deerfield's flaws and then they gazed forward, knowing their next baby would be better-looking and smarter, too.
>
> The man-made North and South Lakes went up, marking the top and bottom of the lasso known as the Yale Loop. This is the continuous circle of a street with amazingly wide lanes and grassy dividers and airy trees placed at equal intervals (they stay deliberately green in the winter and sprout purple blossoms in spring and summer). Everyone always says that nature has an incredible sense of balance, but they need to come look around here. One white gazebo at the South Lake? One white gazebo at the North Lake. One faux-rickety bridge (hence, the name) at the South Lake? "Aw shit, I got that, too" laughs the North. A "lagoon" with concrete rock slides and imported sand? Both ends of Woodbridge raise their hands and then fight over which one has less duck crap in it.
>
> And so it goes. Absolutely nothing has been left to chance. I think Irvine even let Costa Mesa and Santa Ana pop up next to it on purpose, so it would have somewhere to put its maids and gardeners …
>
> On my way home I walk past the South Lake, where citizens can rent small boats and drift around if they feel like it (don't worry, the North Lake has got all this, too.) Today there wasn't a single boat out. In this town, if you build it, they won't come. Ha. Or actually, I think it's just that the people here recognize the lake for what it really is, which is a sculpture of a lake instead of an actual living body of water (Figure 4).

Figure 3. Like the Red Panda *(2004)*

Figure 4. South Lake in Woodbridge Village

Several other new towns such as Mission Viejo, Aliso Viejo, and Rancho Santa Margarita, which SOJA (1996, 113) derides as "super-dormitories" or "cities that are doubles of themselves," were conceived and executed in a like manner, from ranchland to master-planned. In the area now known as "South County," eight cities were incorporated between 1988 and 2001, "all of them indistinguishable from one another" (ARELLANO 2008, 58).

In addition to its planned communities, southern Orange County is the epitome of a postindustrial economy comprised of financial services, real estate, and information (KLING et al. 1991). Irvine, Costa Mesa, and Newport Beach each boast impressively large and decadent office parks mixed with shopping malls and various other cultural establishments: the North and South Koll Centers, Jamboree Center, Newport Center, South Coast Metro Center, Town Center, Civic Plaza, and South Coast Plaza, to name but a few (SOJA 1996) (Figures 5 and 6). For all these downtown-inspired, several-storied centers and plazas, Orange County is not exactly urban, nor is it exactly suburban. Its form is aptly categorized by KLING et al. (1991, 3) as "postsuburban." It lacks the traditional suburban dependence on a larger city for jobs and cultural activities (what might have been its relationship to LA), and although categorized by the Census as a Metropolitan Statistical Area, it has no dominant city and no identifiable core, as do more traditional urban areas. Instead, it is, in SOJA's (1996, 238) words, "Exopolis," an ambiguous place of "city-full non-cityness," a multi-centered, yet center-less conurbation.

Figure 5. Irvine's Jamboree Center

Figure 6. Costa Mesa's Town Center

The spatial organization of South County, granting ample space and prominence to various retail outlets, does nothing to hide what KLING et al. (1991) identify as the region's principal cultural trait, consumerism. "What is socially important in Orange County," explains MACCANNELL (1992, 77), "is not actual values, but the public expression of inflated values." Consumption is indeed undertaken conspicuously here: Orange County is home to the largest grossing shopping mall in the country (South Coast Plaza), the world's largest Mercedes-Benz dealer, the most expensive auto dealership ever built (Newport Lexus) (Figure 7), the second largest BMW dealership in the U.S., the most purchasers of Lamborghinis in America and the second-most of Aston-Martins (ARELLANO 2008). Its residents hold among the most purchasing power in the nation (VENKATESH 1991), the county contains the most expensive housing market in the U.S. (Newport Beach) (Coldwell Banker 2011), and a developer claims that his showy Orange County oceanfront abodes inspired the term "McMansion," or the millionaire version of Orange County's signature tract home (BROOKS 2001, ARELLANO 2008) (Figure 8). All boast "four-car 'garage mahals' jutting out front, altarlike spas in the master baths," and an impressive-looking high-ceilinged space just inside the front door, where the house achieves its all-important "impact on entry" (BROOKS 2001). As described in the Lonely Planet Guide to Coastal California (VLAHIDES 2007, 207), "... materialism, rather than conservatism, does seem to be the OC's one unifying sentiment" (Figure 9).

Figure 7. South Coast Plaza

Figure 8. Newport Lexus

Figure 9. Newport Beach McMansions

Orange County is not just affluent, but the level of comfort achieved by so many of its residents led to its role as the birthplace of modern conservatism. McGirr (2001) explains that this distinction is a consequence of Orange County's particular style and pace of growth. Although the region's early growth was dependent upon government defense contracts, many long time residents embraced a typically Western "ethos of individualism" and resented government control, even though decisions made in Washington had made them rich (McGirr 2001, 30). The stunning economic success achieved by native Orange Countians only strengthened existing libertarian tendencies. Migrants to the area also did well:

> They arrived poor but became, if not wealthy in every case, at least middle class and better off than the Mexican Americans with whom they share the region. Theirs is not a Philadelphia, Mainline, or New York, Wall Street-type conservatism based on clear class divisions and the preservation of traditional privilege. It is a petit bourgeois conservatism based on trivial distinctions, hair color, automobile 'detailing' ... or a few thousand dollars' difference in the average price of the same model home in adjacent tracts (MacCannell 1992, 76–77).

Settlers in the sometimes gated, über-planned new towns were welcomed by a built environment which effectively privatized public space and encouraged individualistic concerns of maintaining property values and "keeping up with the Joneses," rather than directing their energies toward the good of the greater community (McGirr 2001). Residential development in southern Orange County, in MacCannell's (1992, 78) words, "is the most perfect expression of conservative values ... a monument to individualism, private initiative, and 'free-dom.'" New migrants, largely Midwestern and Bible Belt stock already possessed of right-of-center Protestant values, adapted to this new form of community, or lack thereof, by adopting or intensifying conservative tendencies to protect their piece of the American Dream (McGirr 2001).

As a reaction to the striking social changes of the 1960s and 70s and in search of a sense of community deficient in this Orange County lifestyle, residents flocked to the region's growing evangelical churches. Orange County is the home of Reverend Chuck Smith's far-reaching Calvary Chapel, one of the largest Christian ministries in the 1970s, which held mass baptisms in the Pacific Ocean, as well as the Trinity Broadcasting Network, which played an important role in popularizing televangelism (Arellano 2008) (Figure 10). Orange County is also where *Hour of Power* televangelist Robert Schuller got his start. Schuller's sermons, once shouted from atop a snack bar at the nation's first drive-in church (by nights a drive-in movie theatre), are now broadcasted from the breathtaking Philip Johnson-designed Crystal Cathedral, a "white-steel-and-silver-glass extra-vaganza of visual prayer" (Soja 1996, 248) (Figure 11). This showy brand of worship appealed to Orange Countians particularly, as "Schuller sought to give meaning to lives filled with materialism, while not challenging the grounding of those lives" in conspicuous consumption (McGirr 2001, 253). Schuller assured his flock of their "God-ordained right to be wealthy" (Arellano 2008). More recently, Saddleback Church (founded by Pastor Rick Warren, best-selling author of *The Purpose-Driven Life*), occupying a 120-acre campus in southern Orange County, has become the fourth largest mega-church in the country (Arellano 2008) (Figures 12 and 13).

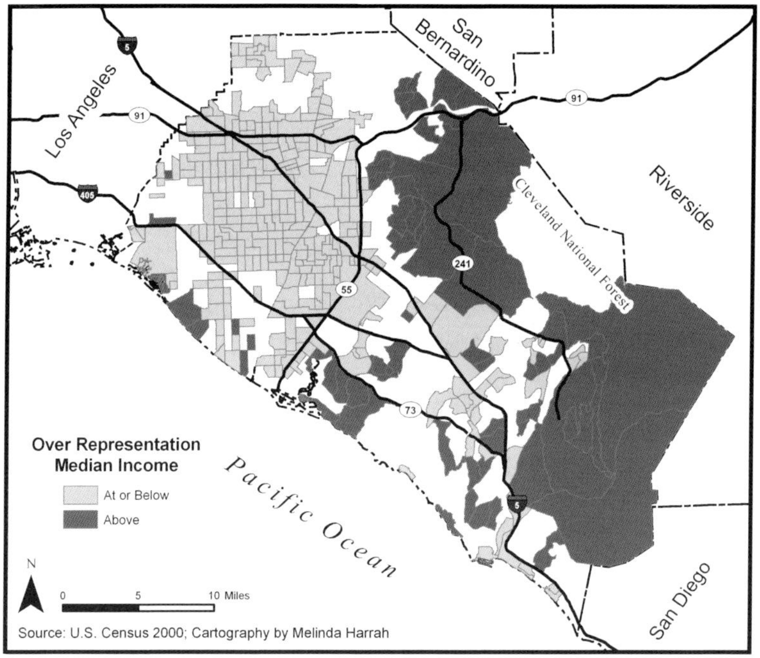

Figure 10. Orange County Census tracts where median income is above the national average

Figure 11. TBN World Headquarters in Costa Mesa (open for tours)

Figure 12. The Crystal Cathedral

Figure 13. Saddleback campus in Lake Forest

The confluence, in Orange County, of many successful fiscal conservatives/libertarians and large numbers prescribing to the values of the Religious Right, created a strong conservative coalition. This unique political environment spawned a grassroots movement which enlisted desperate housewives and soccer-mom prototypes to propel Barry Goldwater to the Republican presidential nomination in 1964 and clinched Ronald Reagan's election as Governor of California in 1966. This once-localized interest group quickly grew to become a national force, culminating in the ascendance of Reagan to the presidency in 1980 (McGIRR 2001). Orange County and its conservative Christian base have retained a strong influence upon national politics. As testament, 2008 presidential nominees Barack Obama and John McCain made their first joint appearance at a forum moderated by Rick Warren at his Saddleback Church, and Warren was selected by President Obama to deliver the invocation at his 2009 inauguration.

THE ANTI-LA

Orange County is often derided by Angelinos as too homogenous, too conservative, too superficial. On the other hand, Orange County has often promoted itself as the anti-LA – less congested, less violent, more controlled. Indeed, Orange County itself was created for the very purpose of distancing residents from the "urban problems" plaguing neighboring Los Angeles, and now the lure of southern Orange County's new development depends on creating a small-town identity which contrasts with the insidious "other" looming to the north (TILL 1993, 720). This "other" can be read as "dangerous, corrupt, and immoral" Los Angeles, or as increasingly smoggy and congested northern Orange County, victim to the same ills as the Inland Empire (TILL 1993, 722). For example, the promotional literature for one of Orange County's new towns, Rancho Santa Margarita, describes its appeal by labeling sprawl as the encroaching enemy:

> For people who like small towns, the choice is getting smaller. Because in this age of growth the small town is disappearing. Merging with cities, and other towns in a phenomenon called sprawl. Urban sprawl. Where can you go to escape ... without totally distancing yourself from all that the large city offers? Today, almost nowhere. Unless you can find a place as gifted by nature as Rancho Santa Margarita (Santa Margarita Company 1986, 5, as cited in TILL 1993, 724).

As TILL (1993, 724) concludes, "the 'American dream' is represented as no longer available in Los Angeles, but is present in the future of Orange County."

Viewed from the other side of the county line, however, Orange County is a rather prudish neighbor. Rather than Hollywood stars or the Sunset Strip, its "chief cultural contribution" is the planned community (MORRISON 1990, 25). Its nickname, the "Orange Curtain," is homage to residents' insulated lifestyle; the County remains a Republican stronghold, and while Los Angeles retains a rather high rate of violent crimes, Irvine is one of the safest cities in the U.S. (DRISCOLL and DOSS 2005). The title character in Sonia SINGH's (2004, 182) novel *Goddess for Hire*, succinctly isolates the difference between the two counties: "Both regions had their plus points. LA had hipness. Orange County had parking" (Figure 14). A 1990 cover story in the *Los Angeles Times Magazine* which investigates the question "LA Versus OC: Just How Different Are They, Anyway?," describes the Angelino's outlook on Orange County:

> At its best, Orange County comes across as well-groomed, generous of creature comforts, pretty in a standardized, debutante fashion, in part from master planning that has become a worldwide model … At its worst, Orange County appears to be nouveau-riche snobbish, insecure, artificially flavored and colored, vapid, priggish and drearily sanitized … but a trifling stop on the way to San Diego (MORRISON 1990, 12–13).

Figure 14. Goddess For Hire *(2004)*

Yet, the article concludes that perhaps Orange County and LA are not so dissimilar, after all. As is the case in several infamous Los Angeles neighborhoods, many Orange County households, mostly in the north, fall below the poverty line (Figure 15). For its all-white reputation, Orange County, since 1993, is one of California's "majority-minority" counties; it is home to the largest Vietnamese population outside of Saigon (in Westminster), its City of Santa Ana (slogan: "Downtown Orange County") is seventy-eight percent Hispanic, and even well-coiffed Irvine boasts an all-Vietnamese Boy Scout troop (U.S. Census 2000; CAMPBELL and PEREZ 2004; BIRD 2008) (Figure 16).

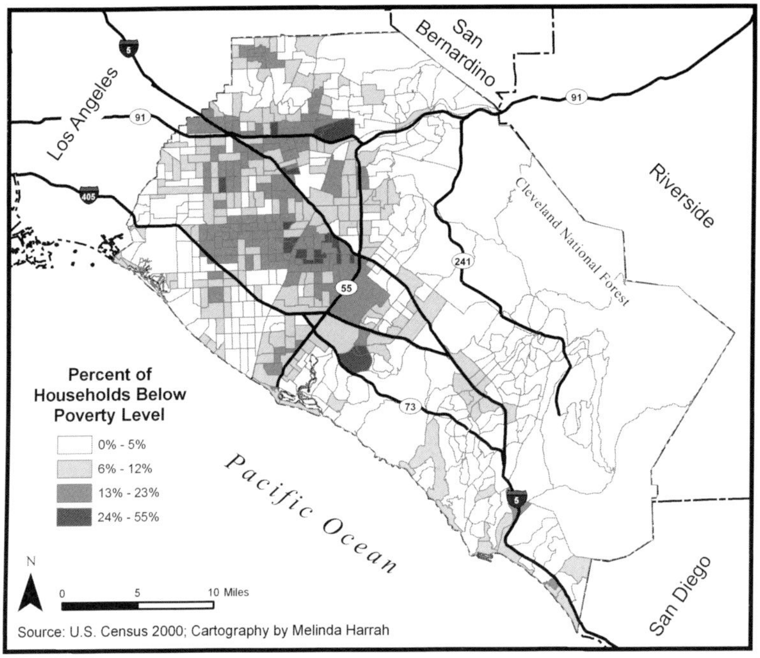

Figure 15. Percent of Orange County households below the poverty line

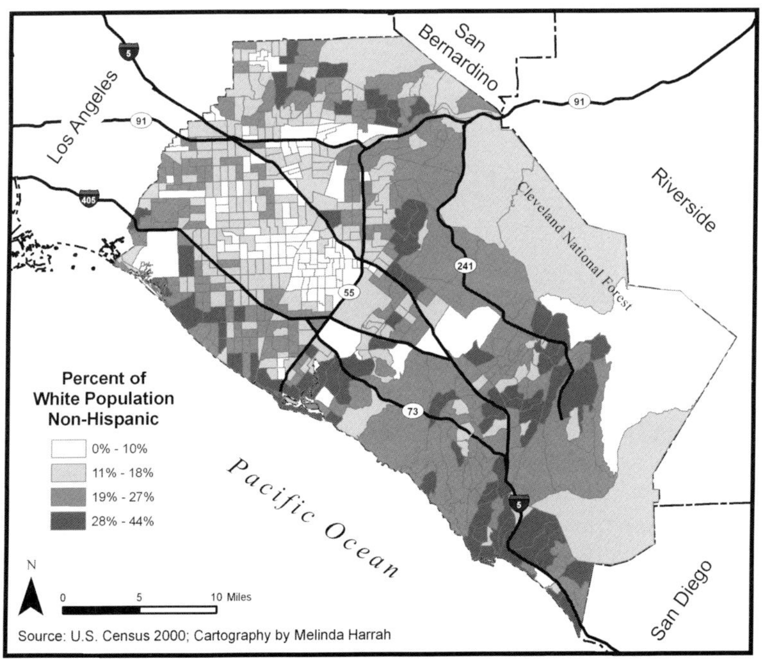

Figure 16. Distribution of the white non-Hispanic population in Orange County

A visitor to Orange County might also be surprised to find more than a few cultural offerings: a large and prestigious university (UC Irvine), several good museums, a number of ethnic restaurants, and across the street from South Coast Plaza, the multi-million dollar Orange County Performing Arts Center, in essence not so different from downtown LA's *Disney* Concert Hall (Figure 17). All this leads Judith Krantz, author of the best-selling romance novel *Scruples*, set in ritzy Beverly Hills, to proclaim: "I feel [Orange County is] a new, hot place, a very, very hot place, from every point of view … It's changed, it's not what it used to be. It's been a place for 100 years, but it's now finally a *place*" (MORRISON 1990, 39).

Figure 17. Segerstrom Hall at the Orange County Performing Arts Center

Both counties were built upon the same image, that sunny Southern California mystique. "Are [they] not warmed by the same ultra-violet rays, quenched by the same pirated water, ensnared in the same gill net of freeways?" asks MORRISON (1990, 10). As compared in the Moon Handbook guide to Coastal California (WEIR 2004, 547):

> Born into endless summer, freed from community by freeways, and taught to believe that here, life can be all things to all people, Orange County and Los Angeles are actually very much alike. But LA is older, more experienced … the City of Angels seems more willing to acknowledge the shadow side of the sunny Southern California dream …

While they have experienced similar boom-and-bust cycles, LA's promise, lost to crime, congestion, and lack of open space, has dwindled more quickly (MORRISON 1990). Although the shrinking frontier may still be for sale in one of its new planned communities, Orange County has a dark side as well.

THE TRAGIC KINGDOM

Similar to LA's depiction as "fallen paradise" in noir films such as *Chinatown* (1974) and *LA Confidential* (1997) (TILL 1993, 722) (Figure 18), merciless ghetto in *Boyz n the Hood* (1991), or bleak setting for unsatisfied yuppies in *Scenes from a Mall* (1991), popular (pre-*The OC*) media representations of Orange County overwhelmingly depict either the American Dream fallen flat in stale suburbia, or a paradise lost to overdevelopment.

Figure 18. LA Confidential *(1997)*

The murder mystery novels of T. Jefferson PARKER are often set in Orange County. His first best-seller, *Laguna Heat* (1985), draws upon a "trouble in paradise" theme: a rare murder has been committed in sleepy Laguna Beach, where, according to the back cover promotion,

> every day the sun makes a promise the nighttime breaks ... Where trouble has swept in like a Santa Ana wind, blowing the cover off a world of torture, murder, and blood-red secrets ... Where a crazed killer has turned paradise into a Disneyland of depraved violence ... (Figure 19).

In *Pacific Beat* (1991), however, the action moves up the coast to Newport Beach, painted by PARKER as a place far from paradise, a community tainted by the greed of developers and of course, murder (Figure 20). In one passage, PARKER's *Pacific Beat* protagonist, Jim Weir, has journeyed inland, and through his eyes we see a ruined Orange County:

> They started off for the coast, threading first through the barrio side streets of Santa Ana where the houses stood behind wrought-iron fencework to keep the thieves out – even the windows were latticed with metal that tried to look decorative but was, in fact, a barrier

against the junkies, crackheads, gangbangers, drifters, home invaders, cutthroats, and occasional killers who wanted in. Then down Fourth Street, past the cafes and shoe stores and taco stands; past the record stores blasting mournful Mexican ballads; past the pawnshops and the beauty parlors; through the crosswalks filled with women in dresses, burdened by groceries, laboring flatfooted from the marketplaces toward home. Men moved even more slowly along the sidewalks, men with cowboy hats and heavy jackets and sun-darkened faces, men without work, without applicable skills or pressing expressions – barely visible beneath their hat brims – of acceptance, resignation, and faith that the Virgin Mary or certain saints would eventually deliver them from this hostile land of dreams mañana, mañana, mañana (PARKER 1991, 168).

Jim guided the truck down Fourth, out of the barrio and into the white suburbs of Tustin, a city that, for years, had actually maintained a sign on this boulevard that read "Welcome to Tustin – the Beverly Hills of Orange County." Jim could remember thirty years ago, when there had been on odd bit of truth in this. There were big stately homes in the foothills, small ranchos tucked behind strands of eucalyptus and avocado overlooking mile upon mile of emerald green citrus groves. The smell of orange blossoms – visceral, opiate – rose on invisible thermals. Main Street looked like a dictionary definition of itself – solid brick buildings housing the five-and-dime, the garage, the pharmacy, and an occasional Victorian manse converted to professional offices. But in the late sixties and early seventies, Tustin had sold out. Franchises now ruled every street corner – the Golden Arches, the Jack in the Box, the garishly unmistakable logos of oil conglomerates, grocery chains, convenience networks, and fast-food distributors – all bawling for consumer attention. It was astonishing to Weir, as he gazed out through the eye-watering smog, that in the city of Orange, just a few miles from here, a special plot of ground had been planted with orange trees so people in the future could see what they really looked like. They had literally paved over the trees and put up a tree museum. At some point, he thought, isn't enough enough? (PARKER 1991, 171).

And yet, thought Weir, the beat goes on. Jackhammers tore out old asphalt to widen the offramps and add freeway lanes; foundation crews and framing teams scurried to cover the last inches of unbuilt land; developers crammed thousands of identical units over hillsides and small valleys while city councils and the Board of Supervisors approved it all from on high with furtive sweeps of the ballpoint. It seemed to Weir to be less the land of dreams than the land of disregard. It's all for sale under the spacious skies, he thought, every last amber wave of it … (PARKER 1991, 171–172).

Figure 19. Laguna Heat *(1985)*

Figure 20. Pacific Beat *(1991)*

In *The Gold Coast*, Kim Stanley ROBINSON (1988) paints a suggestive, if not disturbing, picture of the Orange County that could be (Figure 21). It is the year 2027 and growth has continued unfettered. Orange County is now one of the most densely populated areas in the world. Having run out of room on the ground, development has progressed in multi-layer fashion: "Lower" Santa Ana, confined under a sky of concrete, is a dark and dangerous slum (ROBINSON 1988, 100); high rise condos, apartment complexes, and offices tower cover the County and even Cleveland National Forest has been opened to development; "San Diego, along with Riverside and Los Angeles and Ventura and Santa Barbara, is nothing more than an extension of OC" (ROBINSON 1988, 141).

Figure 21. The Gold Coast *(1988)*

Also included in ROBINSON'S (1988) nihilistic vision: Orange County has become an "autopia" where freeways are stacked hundreds of feet in the air; malls have become so massive that it is possible to live, eat, shop, and exist within them without ever going outdoors; drug addiction is rampant; and only two percent of Orange Countians now attend church. The U.S. is waging war on several countries and the defense industry is incredibly powerful and corrupt. As described by Jim, a young man who longs for orange groves long ago destroyed:

> Orange County is the end of history, its purest product. Civilization kept moving west for thousands of years, in a sunset tropism, until they came to the edge here on the Pacific and they couldn't go any farther. And so they stopped here and did it. And by that time they were in the great late surge of corporate capitalism, so that everything here is purely organized, to buy and sell, every little piece of us (ROBINSON 1998, 3).

No Doubt, a popular ska-rock band from Anaheim, echoes this mixture of disgust and nostalgia on the title track of their hit album, *Tragic Kingdom* (1995), a play, of course, on Disneyland's Magic Kingdom:

> Once was a magical place
> Over time it was lost
> Price increased the cost
> Now the fortune of the kingdom
> Is locked up in its dungeon vaults

Text on their album cover reads "Bought and Sold Out in the USA" (Figure 22).

Figure 22. No Doubt's Tragic Kingdom *album cover*

Other dystopic visions of Orange County express disdain for the suburban life it exemplifies. Philip K. DICK's (1977) science fiction novel, *A Scanner Darkly,* and its recent film adaptation (2006), is set in Anaheim in the not-so-distant future, a bleak period during which a significant portion of the population is addicted to a brain-damaging drug called Substance D (Figure 23). The main character is Bob Arctor, a cop so deep undercover and ruined by Substance D that he can no longer discern his own identity. DICK (1977, 64) describes the moment of Arctor's dissatisfaction with the typical Orange County life he once led, the rejection of which was followed by his entrance into the dark, yet more stimulating, world of a narc/drug dealer:

> There had been a time, once, when he had not lived like this, a .32 under his pillow, a lunatic in the backyard firing off a pistol for God knew what purpose … In former days Bob Arctor had run his affairs differently: there had been a wife much like other wives, two small daughters, a stable household that got swept and cleaned and emptied out daily … But then one day, while lifting out an electric corn popper from under the sink, Arctor had hit his head on the corner of a kitchen cabinet directly above him. The pain, the cut in his scalp, so un-

> expected and undeserved, had for some reason cleared away the cobwebs. It flashed on him instantly that he didn't hate the kitchen cabinet: he hated his wife, his two daughters, his whole house, the backyard with its power mower, the garage, the radiant heating system, the front yard, the fence, the whole fucking place and everyone in it. He wanted a divorce; he wanted to split. And so he had, very soon … Probably he should have regretted his decision. He had not. That life had been one without excitement, with no adventure. It had been too safe. All the elements that made it up were right there before his eyes, and nothing new could ever be expected. It was like, he had once thought, a little plastic boat that would sail on forever, without incident, until it finally sank, which would be a secret relief to all.

Figure 23. A Scanner Darkly *(2006)*

The Orange County environment has produced a large number of disaffected suburban youth, and in the early 1980s they formed an important center of the hardcore punk scene. The "American Hardcore" style of punk rock was very raw and played lightening fast, with lyrics shouted by angry young men. Fueled by distaste for Reagan-era middle class values, Hardcore was the "suburban response to the Punk revolution" (Blush 2001, 12). Orange County kids were drawn to this music, and bands such as Middle Class, Agent Orange, the Adolescents, TSOL, and Social Distortion sprang from the sprawl of Santa Ana, Fullerton, and Huntington Beach (violent slam-dancing was born of the Huntington Beach scene) (Figure 24). Explains Blush (2001, 86): "No artistic hipsters, these were products of the American Dream gone awry, by way of broken families or parental apathy, with some monsters among them." Unlike more artistic, urban-based punk from New York or London, "… in the suburbs like Orange County, it was crazy kids who needed an excuse to go nuts, rebel, and let off steam" (Rikk Agnew as cited in Blush 2001, 86). Casey Royer of The Adolescents recalls, "We were frustrated and trying to find a way out of … basic suburban normality" (as cited in *American Hardcore* 2006). Tony Cadena of The Adolescents explains the appeal of this music to an Orange County teenager:

> Orange County was a very suburban Right Wing, White middle-class stifling environment … We were dysfunctional, a welfare family living in an upward middle-class neighborhood. My family did not fit in. Punk came along at the right time – it gave me a chance to reject everything I couldn't have anyway (as cited in BLUSH 2001, 86).

Mike Ness of Social Distortion says this of his home:

> I love Orange County but it's a love-hate thing. I love to go down to the beach and go surfing but every now and again, I'm reminded about where I live … I need stuff to be discontented with. That's how I write songs. There's still plenty of it left in Orange County, believe me (VH1 2002).

Figure 24. Agent Orange's Living in Darkness *(1981) album cover*

More recently, and in lighter fashion, the 2003 television comedy *Arrested Development* followed the travails of a quirky, entitled Newport Beach family in the development business, now strapped for cash due to the patriarch's mishandling of funds (Figure 25). The opening scene of the series shows members of the family living in a lone model home in a shoddy housing development that is not, to say the least, progressing as expected. Similarly, the 2002 feature film, *Orange County*, chronicles the attempts of high school senior, Shaun, to escape an intellectually-stifling Orange County by gaining admission to Stanford University (Figure 26). The movie opens with a monologue by Shaun, an apt description of the Southern California teen stereotype:

> I live in California in a place called Orange County. A year ago I was just another surfer, spending my days at the beach with my buddies playing volleyball, spending my nights around a bonfire partying … [Then I realized,] maybe there's more to life than extreme sports and trying to get laid … I don't think Orange County is the best place for an aspiring writer so I'm applying to Stanford.

Shaun is hindered in his quest to get into Stanford by a full slate of Orange County caricatures: an inept guidance counselor who mishandles his transcript, a principal more interested in getting Britney Spears to speak at graduation than Toni Morrison, a possibly illiterate English teacher who lectures on movies based on books, an alcoholic mom who married an old man for his money, a wealthy father who married a bimbo, a slacker brother, surfer friends who like to "get lit" and blow things up, and finally, a promiscuous cheerleader who gets into Stanford solely because her grandfather is on the board. At the end of the film, Shaun realizes that Orange County and its cast of characters inspire his writing, and his story proves: "Even in a world where people can be superficial, stupid, and selfish, there's still hope."

Figure 25. Arrested Development *(2003)*

Figure 26. Orange County *(2002)*

The multiple identities associated with Orange County – as site of the American Dream, the anti-LA, the Tragic Kingdom – were soon to be eclipsed by a new image of Orange County presented to Americans on primetime television. Of course it was not the barrios of Santa Ana, the sprawl of Anaheim, or the ethnic streets of Little Saigon which TV took as its subject, but as MACCANNELL (1992, 76) describes, our screens have been filled by "the mythic side" or "the blonde side" of Orange County. Through the presentation of seductive coastal landscapes and the charmed lives of its residents, television has created an Orange County consisting of equal parts sunshine, beauty and wealth. As will be described in the next chapter, the popularity of such shows has fixed this image of Orange County in popular consciousness, not as a boring bastion of suburban conservatism, but as an exceedingly trendy place.

CHAPTER 3

THE CAPITAL OF COOL

Judith Krantz' 1990 proclamation of Orange County's arrival as "finally a *place*" (MORRISON 1990, 39) would reach fruition more than a decade later, with a string of positive publicity touting the county as a national trendsetter. In 2002, major cable network VH1 produced a documentary titled "Orange County: American Hip Factory," which, in asking the question "Has this idyllic suburb become the new Seattle?," recognized Orange County's important contribution to the contemporary rock music scene in the bands Social Distortion, No Doubt, the Offspring, and Sugar Ray (VH1 2002). Later in 2002, *USA Today* dubbed Orange County "the new capital of cool" on the basis of its cutting-edge shopping malls (The Lab Anti-Mall, The Camp), its local big-name surf and skatewear companies (Quiksilver, Oakley, Vans, Hurley, Volcom), and even its "cool grub" (the chain Wahoo's Fish Tacos) (HOROVITZ 2002). Following suit, *U.S. News and World Report* ran the headline "Hip's New Headquarters," also in reference to Orange County (STREISAND 2003). The somewhat successful adolescent comedy film *Orange County* had also debuted in 2002, under the tagline "It's not just a place, it's a state of mind." Summarizing these developments, the *Los Angeles Times* offered this comparison of new versus old OC (YOSHINO 2003):

OLD	NEW
Eating fried chicken at Knott's	Eating pho in Little Saigon
Righteous Brothers	The Offspring
John Birch Society	Surfrider Foundation
Gene Autry	Arte Moreno
Crystal Cathedral	Saddleback Church
Santa Ana Fashion Square	Irvine Spectrum
Nixon's Western White House	Dennis Rodman's beach house in Newport
Leisure World	Santa Ana Artists Village
Carl's Jr.	Wahoo's
Santa Ana Freeway	San Joaquin Hills Toll Road

THE OC

Then, in August 2003, a television show called *The OC* debuted on Fox (Figure 27). A natural conduit for the county's swelling pop culture presence, TV would broadcast its hipness to a much larger audience. Trailers for the show warned those viewers expecting a frolic in suburban sameness: "It's nothing like where you live. And nothing like what you imagine." In contrast to *Orange County* the movie, in which its setting served only as background, doing little "to show what exactly constituted the place or the state of mind" (ARELLANO 2008, 217), *The OC* presented Orange County in a manner deserving of a title role. Set in upscale Newport Beach, the show introduced viewers to the sun-drenched exclusivity of "the OC" through the journey of a troubled youth, Ryan, from the wrong side of the tracks in rough Chino, California. Ryan is taken in by a wealthy Newport family, the Cohens, and the show is premised on the struggles of Ryan and the Cohens' awkward son, Seth, to fit in with the superficiality of mainstream Newport society. Each week, the show featured dazzling Orange County coastlines, tanned teens enjoying watersports, and extravagant wealth, set to a California-evoking indie soundtrack.

Figure 27. Cast of The OC *(2003)*

The OC is best described as a "primetime soap," crafted in the vein of 1990s super-hit *Beverly Hills 90210*. Also set in a very wealthy area, Fox's *90210* was the first primetime high school drama, ground-breaking for its portrayal of young people dealing with adult issues, not to mention driving expensive cars (ITZKOFF 2008); its appeal lay in its depiction of "high school life not as it was experienced but rather how teenagers dreamed it ought to be" (BELLAFANTE 2007) (Figure 28). *90210* ended its ten-year run in 2000 and Fox producers found their follow-up in *The OC* (SCHWARTZ 2004).

Figure 28. Cast of Beverly Hills 90210 *(1990)*

In addition to Ryan and Seth, *The OC*'s core cast includes best friends Marissa and Summer, and Seth's parents, Sandy and Kirsten Cohen. Kirsten is the daughter of a wealthy OC real estate magnate and holds a high position in his development company. Sandy, a Jew from New York, is a public defender, who would not, apart from his wife's position, be living in Newport Beach. The two met, of course, in college at Berkeley during more idealistic times. Marissa is a very pretty but troubled neighbor whose mother is obsessed with appearances and whose father is a stockbroker (secretly) going bankrupt.

The first season unfolds as follows: tough but sweet Ryan comes to town, shows his knack for getting into fights, offends/threatens the high school jocks, he and Marissa fall in love. Marissa's parents get divorced, she gets into some trouble in Tijuana. Ryan draws Seth out of his shell, Seth succeeds in wooing Summer. The season ends with these cliffhangers: Ryan is led to believe his ex-girlfriend in Chino is pregnant so leaves Marissa and Newport to "do the right thing." Kirsten has, in spite of herself, grown quite attached to Ryan and cries. Seth is devastated by Ryan's departure and sets sail into the Pacific, Marissa's mother marries Kirsten's father, and Marissa, also devastated by Ryan's departure and depressed by her mom's remarriage, drinks heavily. All of this presented in montage to the tune of Jeff Buckley's (OC native) haunting rendition of "Hallelujah" is, needless to say, a heart-wrenching conclusion to a nail-biting season (Figure 29).

Figure 29. Ryan leaves Newport in the season one finale

The OC's first season was very successful, drawing nearly ten million viewers per episode (WYATT 2007). Much of its early success can be attributed to the vision of its creator, Josh SCHWARTZ, one of the youngest persons ever to oversee a primetime series. His idea to set a show in Orange County stemmed from his experience attending the University of Southern California after growing up in Rhode Island:

> I got out there and was sort of overwhelmed by these Nordic-blonde water polo-playing guys and their beautiful girlfriends … I got to spend some time with them and go into their gated communities and their McMansion developments, and their life on the surface seemed perfect … (SCHWARTZ 2004).

After its fast start, *The OC* met a rather swift demise. Viewership held fairly steady through season two and dropped somewhat in season three, which ended with Marissa's death. Without her popular character, and placed in a competitive Thursday night time slot, viewership shrank to barely four million in *The OC*'s fourth and would-be final season (WYATT 2007). The show's finale ended as the Cohens moved to Berkeley, and Orange County would never be the same. Its image had been transformed, in the words of one columnist, "from Los Angeles' ho-hum neighbor to a trendmaker perched on the endless Pacific" (VILLARREAL and POWERS 2006). Another observes:

> Particularly since *The OC* … became a pop-culture phenomenon, the formerly quiet patchwork of planned communities and insular beach towns … has become recognized worldwide, however improbably, as the embodiment of all things hip, youthful, and glamorous. To those of us with roots in Orange County, that is bizarre. It is as if the organizers of New York Fashion Week had pulled up stakes in Bryant Park and relocated their tents to Bergen, New Jersey, the newly discovered capital of cool (WILLIAMS 2004).

LAGUNA BEACH: THE REAL ORANGE COUNTY

The popularity of *The OC* and its appealing depiction of a privileged Orange County lifestyle led MTV to develop a reality show based on real teens living therein. When Liz Gateley, the show's creator, pitched the concept to MTV executives, the question was posed: "Can you tap into a clique of friends and have them all be interesting and beautiful?" A native of Palos Verdes, she answered "They're a dime-a-dozen in Southern California. They're everywhere" (ROCHLIN 2005). Thus, in 2004, *Laguna Beach: The Real Orange County* became the first of a new subgenre: the "reality drama" (CHANG 2004) (Figure 30). *Laguna Beach* did away with many of reality television's conventions, such as interviews and confessionals, and barred from filming in the local high school, it is somewhat short on action-packed plotlines. So the show found its "drama" in the amplification of minutia: in gossipy trips to the nail salon, the difficulty of picking out a swimsuit for spring break in Cabo, and perhaps in some coaching from MTV producers. *Laguna Beach* quickly became one of MTV's biggest hits, drawing an average audience of three million viewers each week – impressive numbers for a cable network (ROCHLIN 2005).

Figure 30. Season one cast of Laguna Beach *(2004)*

Upon conceiving the show, MTV producers had hoped to film in Newport Beach, the setting of *The OC*, but the city was not inclined to welcome cameras, so they settled instead for Laguna Beach (OVERLEY and CHANG 2007). Laguna Beach is one of Orange County's most picturesque beach communities, and historically, one of its most liberal (Figure 31). Laguna is something of an art colony; its main street contains several galleries and one-of-a-kind-boutiques. Its biggest tourist draws have heretofore been art festivals – the Sawdust Art Festival and the Pageant of the Masters in which people pose in re-creation of famous works of art (the latter event was hilariously lampooned in *Arrested Development*) (ARELLANO 2008). In addition to its artists, Laguna Beach is also home to an active gay community and retains a bohemian charm. As ARELLANO (2008, 221) explains, however, Laguna's housing prices exploded in the mid-1990s and Laguna is becoming more and more like the rest of coastal Orange County – "superrich" and less liberal. With a median value of almost $1.4 million, in 2004 Laguna represented the third priciest housing market in California, ahead of even ritzy Newport Beach (WILLIAMS 2004). Indicative of this shift in sensibilities, in 2002 Laguna Beach High School students voted to change the team nickname from the Artists (with painter's palette and brush as symbol) to the much cooler Breakers (HARVEY 2002; ARELLANO 2008) (Figure 32).

Figure 31. Scenic Laguna Beach

Figure 32. The Laguna Beach Breakers

Laguna Beach: The Real Orange County unsurprisingly depicts a city that is more Breakers than Artists. The principal character and narrator of the series is Lauren Conrad, the daughter of a wealthy architect whose family is in the process of moving to a spectacular new mansion, high above most of Laguna, complete with film-friendly backyard infinity hot tub and pool. Lauren has a crush on Stephen, her long-time friend, who, of course, surfs, skateboards, etc. Lauren's arch nemesis is Kristin, man-eater and Stephen's sometimes girlfriend (Figure 33). Much of the first season's drama revolves around Lauren's attempts to win Stephen's heart. Other notable season one cast members are Morgan, a Mormon with "values" and Christina Schuller, granddaughter of the Crystal Cathedral's Robert Schuller (one episode featured a church service there).

Figure 33. Stephen and Kristin

The love triangle formula is repeated in season two, featuring last year's juniors, now seniors. It is narrated by Kristin, but also checks in on Lauren and Stephen's attempts to adapt to college life in San Francisco. Season two lacked season one's morality; this was *Laguna*'s most popular year. After its stars graduated and mostly moved to LA in misguided pursuit of acting careers, MTV attempted to introduce an entirely new cast in season three, but the formulaic interactions, play to stereotypes, and aura of entitlement which comes with anticipation of great fame had the effect of stripping cast members of personality. In short, *Laguna Beach* "lost its freshness because its cast learned how to portray teenagers by watching *Laguna Beach*" (ABCARIAN 2006). MTV attempted to revitalize its *Real Orange County* franchise by moving up the coast to Newport Beach in 2007, but *Newport Harbor: The Real Orange County* lasted only one short season.

Nevertheless, like *The OC, Laguna Beach* hit a cultural nerve and played a strong role in the solidification of Orange County's place in popular, teen culture. Says one reviewer: "*Laguna Beach* … is the sort of place that people who have never been to Southern California imagine it should look like" (STEINHAUER 2006). "*Laguna Beach* is an oceanfront Neverland where all the Lost Boys and Girls drive SUVs" (GORDON 2005, 64). Another describes the Orange County of these shows as looking like "some landlocked Midwesterner's dream vision of Southern California" (GALLO 2008). Another writes:

> The current fascination with Orange County taps the need Americans seem to have to create a bubble in which playful adolescent fantasies can act themselves out in the sun, removed from any real-world complications … (WILLIAMS 2004).

The executive producer of *Laguna Beach* and *Newport Harbor* offered his opinion:

> … people [in the Midwest] are just crazy watching kids talk and eat in Southern California. And surf and skateboard … LA is its own animal, whereas now Orange County is what everybody in the Midwest and worldwide see as true California (SCHEIBAL 2008).

Both *The OC* and *Laguna Beach* have been criticized for all-white casts largely devoid of color. Although nowhere near representative of Orange County as a whole, as might be inferred from the titles of these shows, they are not so far from demographic realities along the "California Riviera." Newport and Laguna are both 80% to 90% white (U.S. Census 2000). *The OC*, admits a Latino columnist from Orange County,

> portrayed a truthful comedy of errors … Latino roles were limited to hot gardeners, subservient maids, and a pregnant girlfriend … Newport Beach *is* that moneyed and aggressively ignorant about the rest of Orange County (ARELLANO 2008, 219).

THE REAL HOUSEWIVES OF ORANGE COUNTY

Drawing further upon the popularity of Orange County as a setting, *The Real Housewives of Orange County* debuted on Bravo in 2006 and draws approximately one million viewers per episode (SMITH 2007). The show follows five wealthy women who live in an upscale unincorporated gated community, Coto de Caza, in southeastern Orange County (Figure 34). It is touted as "Laguna Beach for grown-ups!" on the cover of its first season DVD.

Figure 34. Season one cast of The Real Housewives of Orange County *(2006)*

The Real Housewives uses a more conventional reality show format, has a lower production budget, and much less scenery than *Laguna Beach*, necessary partly due to the fact that Coto resides less spectacularly inland. *The Real Housewives* was actually pitched as a satire of life in a gated community and titled "Behind the Gates," but the network chose to rename the show in order to benefit from Orange County's newly found cultural cachet (DUNLOP 2008). The title is also meant to provide association with ABC's popular dramatic series *Desperate Housewives.*

The ladies of Coto, by virtue of the town's gates and master planning, actually represent a way of life more common to the Orange County experience than residents of a more organic beach city (although Coto too is about 90% white), but Coto is by no means less affluent. It is one of OC's most exclusive planned communities, a 5,000 acre spread in the "back country" abutting Cleveland National Forest, boasting two golf courses, several horseback riding trails, and even a real orange grove (SMITH 2006). Homes range from golf course condos to your standard 4,000 square foot McMansions to "real mansions with acreage, guest houses, horse stables, moats" (MICKADEIT 2006a). Coto de Caza, in other words, is "an attempt by the superwealthy to segregate themselves from the merely wealthy" (ARELLANO 2008, 222) (Figure 35).

Figure 35. Entrance to Coto de Caza

The Real Housewives still borders upon satire, highlighting the ladies' spending and drinking habits and various quests to defy the aging process. In short, they are presented as "melon-breasted, jewel-encrusted, SUV-driving, margarita-sipping mansionistas" (SMITH 2006). Each episode opens with blurbs spoken by each housewife intended to summarize their persona on the series and their take on the Orange County lifestyle. First we hear Lauri, who has been through several divorces: "I was poor, I was rich, I was poor again, and you know what? Having money is easier." Next, Jeana, a former playboy model and now successful realtor states: "I have always wanted things. I crave money." Vicki, an over-the-top insurance agent chirps: "Here's to not being fake" as we see a shot of her getting eyelash extensions (Figure 36). (Regular viewers know that she has also undergone botox treatments.) Finally, artificially buxom Kimberly informs us: "Eighty-five percent of the women around here have had breast implants." Other memorable quotations from the show's first season (compiled by MICKADEIT 2006b) include:

Lauri: I will never look a day older than 32.
Kimberly: Out here, it is very common that girls receive breast implants for high school graduation.

Jeana: My husband and his mother picked me out of several of his girlfriends because they thought I have the right build for their genetics.

Vicki: If [my son's] friends think I'm a [mom they'd love to have sex with], then great!

Figure 36. Real housewife Vicki

The Real Housewives has, of course, been lampooned by critics and also lacks the "look" and charm of the nostalgic high school dramas. A critic describes the series as "an astute exercise in anthropology dependent on neither gimmick or guile" and goes on to set the scene:

> This being Orange County – specifically the gated community of Coto de Caza, where no cliché of Californian indulgence is ever averted, and a three-car garage just means you're stingy – Vicki languishes in the hot tub drinking white wine with a few of her girlfriends (BELLAFANTE 2006).

Another reviewer writes that the show "is trashier and more believably grotesque" than either *Desperate Housewives* or *Laguna Beach* and "posits Orange County as a vacuous paradise found ... built either to nauseate or fascinate ... which is probably just what the producers intend" (BROWNFIELD 2006). Or, as a Bravo executive more upbeatly describes it, "*The Real Housewives* is one of those great combinations of an aspirational lifestyle mixed with relatable human drama – all creating a guilty pleasure" (CHANG 2007).

Viewers do see that life behind the gates isn't necessarily a walk in the park. Lauri begins the series living in a condo in neighboring Ladera Ranch (she lived in Coto before the divorce), sleeping on sheets purchased from Target (SCHOENKOPF 2008). Jo, a younger "kept" woman is bored and longs for the independence of her own career. Kimberly battles skin cancer and must move back to Chicago. Jeana's husband, a former Oakland A, is absent and emotionally unavailable; in the third season they separate. Vicki is a controlling workaholic in danger of alienating her husband. Tammy, introduced in season two, has also dealt with divorce and downsizing, as well as, in season three, the death of her ex-husband. The children of the housewives are largely spoiled.

The show doesn't stray long from its main agenda, however, which is to capture that "nebulous" Orange County lifestyle (ARELLANO 2008, 223). Along with the heartbreak comes plenty of spending – jewelery-shopping, spa trips, cocktail parties, Mexican cruises and nights out on the town. It would seem that the show succeeds in its efforts to present Orange County as the epicenter of an insular, advantaged lifestyle. If, as *OC Weekly* columnist Commie Girl argues (ironically quoting a character from *Desperate Housewives)*, "perception is reality," then "*The Real Housewives* introduce us … these people are the true Orange County" (SCHOENKOPF 2008, 8–9). Agrees ARELLANO (2008, 221), "It took *The Real Housewives of Orange County* to focus the camera on the people who really make the OC the OC – the bitches."

THE NEW ORANGE COUNTY

Whether viewed as good entertainment or fuel for voyeuristic tendencies, taken together, *The OC, Laguna Beach: The Real Orange County*, and *The Real Housewives of Orange County* present a place that is a far cry from the stale suburbia that was twentieth century Orange County. Although *The OC* is fictional, the *Laguna Beach* kids are probably acting, and the housewives may not be its hippest residents, these shows present an Orange County that is, depending on one's point of view, overdramatic and ostentatious or visually stunning and glamorous. In any case, it is not forgettable. Courtesy of the small screen, Orange County's image has undergone a dramatic transformation from really boring to ultra cool. Television has produced a powerful simulation of Orange County, one that has taken hold of the U.S. consciousness and the popular imagination, such that Orange County is now a conspicuously mediated place.

Through this and the previous chapter, we have chronicled representations of Orange County in popular media, from suburban development and dissatisfaction to plastic paradise and envy of the Midwest. Even before *The OC*, Orange County was a mediated place. As exotic orange groves symbolized the area's promise at the turn of the century, beaches and bikinis are now the main source of its enticing images. Place-making is a social process; each depiction in the media adds to this "meeting place." This chapter has established television as the crux of Orange County's most recent image makeover, and as a powerful agent in the creation of place. The next chapter will examine how television uses geographic realism to construct "the OC."

CHAPTER 4

LANDSCAPE AND PLACE-MAKING ON TELEVISION

How do the television shows *The OC*, *Laguna Beach: The Real Orange County*, and *The Real Housewives of Orange County* use production techniques to assemble the place called "the OC"? In this chapter we argue that the work of place-making occurs through construction of a montage of the four functions landscape serves in television and cinema (place, space, spectacle, and metaphor).

One of the most intriguing tensions in landscape studies is how it is constituted. For some it is "a portion of the earth's surface which can be comprehended at a glance" (JACKSON 1984, 8), or "our unwitting autobiography, reflecting our tastes, our values, our aspirations, and even our fears, in tangible, visible form" (LEWIS 1979, 12). From these perspectives, landscape is a fixed, solid entity, "a text already written" (CRESSWELL 2003, 27), but a text that is nonetheless structured by vision. These characterizations of landscape rely on the belief that it is a cultural product (ROBERTSON and RICHARDS 2003), or a particular "way of seeing" (COSGROVE 1984, 1). But as MITCHELL (2008, 44) notes, as

> a particular, structured way of seeing, landscape has historically been established as *the* way of seeing. Landscape is didactic in that it teaches us to look in certain ways and to value aesthetics (over any number of other ways of knowing) as a means toward understanding the nature, status, and meaning of a place.

Thus, the dominant way of seeing our "unwitting autobiography" (LEWIS 1979, 12) reflects the status quo, but is always in competition "for organizational power within any culture" (FLECKENSTEIN 2006, 70).

As a cultural product, landscape is always-already a representation. Whether seen on the television screen or as a material manifestation, landscape is a "potent cultural symbol" to be consumed by viewers (MITCHELL 1994, 15). If place is a way of understanding the world (CRESSWELL 2004) and a center of felt value (TUAN 1977), then landscape, as cultural product, contributes to place-making by naturalizing meaning, values, and power relations. According to DUNCAN and LEY (1993), naturalizing refers to the ways in which landscape and representations seek to pass off that which is cultural (meaning, values, and power relations) as natural. Those with the power to define the view imbue the landscape with cultural meaning to invoke a particular message about place, one that resonates with viewers. However, while landscapes (and television shows) are sites of cultural production and reproduction, "not all audiences will be able or willing to respond to the way of seeing invited by a particular image and its particular practices of display" (ROSE 2001, 15).

Understanding landscape as a cultural product risks overlooking its dynamism. Landscape is also a cultural process, "an ongoing *relationship* between people and place" (MITCHELL 2000, 102). As we go about our daily lives, we do so within the landscape, continually shaping and re-creating it. In this sense, what matters is "not what the landscape means but, rather, how it comes to be meaningful" (ROSE 2002, 456) and relevant for our own lives. As MCHUGH (2009, 209) notes, "Landscape is emergent as relational lines of movement, an ongoing meshwork of practices and movement signatures." The process of "living" the landscape is the same process inherent in place-making. As landscape becomes meaningful through practice and engagement with it, the place that landscape embodies becomes present – is created. According to MCHUGH (2009, 209), "we are nothing more and nothing less than agents, next selves, 'passing' through. The collective traces of our 'passings' constitutes the making and remaking of place." In the words of CASEY (2002, 271), "Landscapes are, in the final analysis, placescapes; they are congeries of place in the fullest experiential and represented sense." As congeries of place, landscape is co-constituted by the actual (present) and the virtual (past) where memory is vital, "pushing forward in duration, the dynamic continuation of movement and sensation" (MCHUGH 2009, 209). According to BRUNO (2007, 39),

> in this vein, a cultural landscape, broadly conceived, can be regarded in many ways as a trace of the memories, the attention, and the imagination of that inhabitant – passengers who have traversed it at different times. It is an intertextual terrain of passage carrying its own presentation in the threads of its fabric, weaving it on intersecting screens. A palpable imprint is left in this moving landscape: in its folds, gaps, and layers, the geography of cinema ... holds remnants of what has projected onto it at every *transito,* including *emotions*. Imaged in this way, such a landscape is an archaeology of the present.

THE FUNCTIONS OF LANDSCAPE

Landscape, as both cultural process and product, works to create place. Drawing from HIGSON (1984, 1987), LUKINBEAL (2005) asserts that landscape serves four functions in cinema and television: as space, place, spectacle, and metaphor. As space, landscape is a site of dialogue or action, where the drama of the narrative comes to the foreground (AITKEN and ZONN 1994; LUKINBEAL 2005). Landscape as place promotes geographic realism, imbuing television with a specific sense of place. Geographic realism is an aesthetic practice that links a fictional narrative to a sense of place, regardless of whether or not the actual filmed location has any relation to the fictional narrative (LUKINBEAL 2006). Once geographic realism is established, viewers are more apt to suspend their disbelief and accept the meanings appropriated onto the landscape (LUKINBEAL 2005, 2006). HIGSON (1984, 3) explains:

> Landscape and townscape shots must at one level construct a narrative space in which the protagonists of the drama can perform the various actions of the plot. But more than a neutral space, these location shots demand to be read as a real historical *place*, authenticating the fiction.

HIGSON (1984, 9) continues: "place becomes a 'sign of reality': the implication is that it speaks a history, a memory." Memory, movement, and emotion are experienced as a *musée imaginaire* (MALRAUX 1967), a metacinema (MCHUGH 2009), an "emotional map … we experience, on topophilic grounds, an architecture of inner voyage, a geography of intimate space. Filmic site-seeing is immersed in the geopsychic act of interfacing affect and place …" (BRUNO 2007, 39).

Landscape as place (cultural product) is achieved most efficiently through the repetitive use of landscape images in establishing shots. These create a cognitive map in viewers' minds, setting the scene within which narrative space can unfold (LUKINBEAL 2005). Oddly enough, while establishing shots are "tools for spatial orientation," they are almost always montage sequences, a fragmented collection of images (BANERJEE and MARX 2008, 156). According to BANERJEE and MARX (2008, 156), establishing shots fulfill a paradoxical task:

> While functioning as indices of reality for the fictional construction, at the same time they increasingly become elements of a specific cultural code that marks the genre, and to a certain degree, the specific television series.

Repetitive use of images in establishing shots build a representational legacy, a lexicon of icons that "operate as a sort of short hand; they must carry with them a whole baggage of cultural meanings that can be instantly unpacked by the viewer who recognizes the baggage" (HIGSON 1987, 11). Especially in television, operating under the constraints of a timeslot frequently interrupted by commercial breaks, place must be expressed in an efficient manner. SADLER and HASKINS (2005, 201) explain the effect:

> In New York City, the repeated use of the skyline image has reinforced the city's place and has clarified the image in people's minds to the point that most people can recognize the Big Apple from a small piece of the overall image.

Landscape often functions simultaneously as place and spectacle. As spectacle, landscape is presented as "a visually pleasurable lure to the spectator's eye" (HIGSON 1984, 3), an image to be consumed by the viewer, who holds a "position of visual mastery" (HIGSON 1984, 19). Spectacular landscapes suspend narrative space, allowing the spectacle to come to the foreground (AITKEN and ZONN 1994). This suspension of narrative space creates a provocative tension between the desire to linger on the view and the forward movement of the narrative, or the drive to find out what happens next (HIGSON 1984). Narrative instills a "sense of something lacking … a desire to explore," while a spectacular landscape confronts viewers with "an image which is so fascinating that it seems complete; no longer the desire to move on, no longer the sense of something lacking …" (HIGSON 1984, 3). Spectacle, therefore, works on two levels: it constructs emotional maps (BRUNO 2007) that combines (e)motion with memory and intimate space, and it encodes power relations within the gaze, such as the hegemony of the masculine gaze in landscape studies (ROSE 1993) and cinema (MULVEY 1975).

Through the use of metaphor, landscape is infused with cultural or social meaning which links icons and stereotypes to place and spaces (LUKINBEAL 2005). These stereotypes become naturalized ways of seeing, and like icons, they are constructed and reinforced through repetition. For example, in cinema, the city is often presented in opposition to the country to represent loss of innocence (MCARTHUR 1997). Metaphorical landscape can visually express a character's state of mind, as rain often signifies sadness: "it is not just that the character is in the landscape, but that the landscape becomes part of the character" (HIGSON 1984, 12). Landscapes of metaphor and spectacle can both generate pauses in narrative, producing "meeting places" where memory, emotion, actions, and events craft a representational legacy – a place imbued with meaning to viewers of the show. Metaphor, like spectacle, can also ease the tensions between place and space by allowing the social meaning(s) of place to be appropriated to space and vice versa.

Finally, space is central to the construction of social meaning. HIGSON (1987, 8) describes landscape's primary function on television as a "discourse space: that is, as the space in which people talk … as figures in the landscape." These spaces are integral to the ongoing process of place-making. Within such spaces important dialogue is uttered by characters, revealing facial expressions are captured in close-up, and movements are tracked, all presented within a television show to build characters' personalities and expand viewers' understanding of their lives and situation. Here landscape is primarily a process; a narrative space of actions and dialogue assumes prominence in these scenes; they further the social meaning of place, deepening our understanding of this televisual world.

As HEATH (1981, 31) notes, in cinema the mise-en-scène is constructed as a narrative space, a quattrocento system in that is a "cued construction of space." Rather than a static and temporarily fixed space (like a painting), mobility in and of the image disrupts the pictorial organization of the composition (cf. DOEL and CLARKE 2007). Movement literally animates (or gives life) that forces narrative space to contain "the mobility that could threaten the clarity of vision in a constant renewal of perspective; space becomes place – narrative as the taking place in film …" (HEATH 1981, 36). HEATH (1981, 37) goes on to argue that "what is crucial is the conversion of seen into scene, the holding of signifier on signified; the frame composed, centered, narrated, is the point of that conversion." Mobility thus poses a problem to the construction of narrative space. It de-centers a coherent place by using it up, requiring spatial cues (such as establishing shots, transition shots, editing, and montage) to return clarity to the narrative. By first directing the viewer's attention to the geographic realism of establishing shots, then to involvement in actions and reactions of the characters within discourse space, and later returning to an establishing shot, there is a constant tension between space and place, or between landscape as cultural process and product, within the narrative.

In the end, it is difficult to separate the functions of landscape as they are techniques used in television and film to create place through ongoing tensions and transitions between space and place, a progression driven by TV's production codes and generic formulas. Landscape as place continually falls away to narrative space and back again (cf. AITKEN and ZONN 1994), and landscape as metaphor and spectacle links, overlaps, and releases the tension between place and space. The process of televisual place-making requires constant tension and continuous scalar movements between landscape, place, and space. Where space constructs a localized sense of social place and allows televisual place to become meaningful through the unfolding of narrative events, place grounds the narrative, provides geographic realism, and establishes broad cultural meanings.

AN ICONOGRAPHY OF ORANGE COUNTY

The televisual functions of landscape are well-illustrated in the pilot episode of *The OC*. The pilot is an important episode; its task is to construct the spaces that characters inhabit and build the foundations of place essential to a successful series. Especially for a show aiming to hook viewers on an extravagant lifestyle, the first impression must be captivating. Orange County is masterfully introduced in the first 15 minutes of *The OC*'s pilot episode.

The show opens in dark, dangerous Chino where our hero, Ryan, is shown as a hesitant accomplice to car theft (Figure 37). Ryan gets caught, spends time in juvie, and is thrown out of his mother's house. With nowhere else to turn, he calls Sandy Cohen, his public defender, and Sandy drives him out of gritty, graffiti-ridden Chino and into beautiful Newport Beach. The camera tracks their car as it drives down the coast, past surfers, sunbathers, and kids playing touch football on the beach, in the striking hues of an ocean sunset. The montage ends as the car enters the Cohens' gated community and pulls into their impressive abode (Figure 38). Through Ryan's eyes, and ours (the camera's), there is quite a contrast between life in the run-down landscape of Chino, and the other-worldliness of a sunny, wealthy, seemingly flawless Orange County. Here landscape functions as a metaphor of contrast: Chino serves as the stereotypical ghetto and Newport the beachfront Babylon. Place-making proceeds as much by using landscape to symbolize what the OC is not (or should never be confused with). By painting place in black and white, in definable disparity, the contrast between these landscapes posits Orange County as paradise, the dream world to which Ryan and the viewers are introduced.

Figure 37. Opening scene of The OC*, in Chino*

Figure 38. Ryan's journey from Chino to "the OC"

Following the pilot, each subsequent episode begins with a traditional opening sequence, which introduces the cast and, of course, "the OC." Establishing shots contained in this montage reaffirm the central elements of televisual Orange County by showcasing landscape as spectacle: first a shot of Newport Coast mansions, then waves breaking on the coast, a sailboat on the ocean, a jet skier, the pier, a beautiful bikini-clad girl, a sunset, and finally Ryan's pool house, rapidly shifting from icon to icon, none lingering on the screen quite long enough to quench viewers' curiosity (Figure 39). Such images move across the screen to the tune of Phantom Planet's anthem "California" whose principal lyrics, "California here we come; Right back where we started from," evoke Ryan's journey to Newport Beach. The opening sequence naturalizes the dominant view of southern California as more Newport than Chino, although in the "real" world these cities are only forty miles apart.

Figure 39. The OC *opening sequence*

This opening sequence, presented each week for 109 episodes, continues to reinforce the show's depiction of Orange County as a coastal playground for the privileged. Landscape as spectacle presents evocative images which define both beauty and the viewer's omnipresent viewpoint via the "god-trick" (HARAWAY 1991). Spectacular landscapes attract the gaze of the audience and establish a particular way of seeing. The place-making process is initiated as these images provide the frame through which we are to experience televisual Orange County.

Landscape as place is further used during each episode to indicate changes in the site of the action or the passage of time. These, which we term transitional shots, often depict characters' homes, places of work, hang-outs, or other locales frequently used in the series (Figure 40). Shots with a longer focal length, such as yachts and sailboats in Newport Harbor or extensive aerial views of the coast are intended to represent Newport Beach/Orange County and reinforce a naturalized geographic realism for the show (Figure 41).

Figure 40. Transitional shot of character's home

Figure 41. Transitional shot of Balboa Pier in Newport Beach

The repetitive use of transitional shots during an episode and throughout the series reinforces and reminds viewers of the naturalized meanings inherent in geographic realism. The harbor vistas to which *The OC* so frequently returns prompt viewers to associate Orange County with wealth and warm weather. These landscapes are used as evidence that the fictional narrative is occurring in the actual location. But as LUKINBEAL (2006, 339) argues, geographic realism "is subjective and fluid, not objective and factual." PECKMAN (2004) also notes that realism is not about mimesis; it is an effect codified by conventions. Ironically, with the exception of some of its establishing and transitional shots, *The OC* is filmed not in Orange County, but rather in Los Angeles. This "crime against geography" (LUKINBEAL and ZIMMERMANN 2006, 8) was committed as a result of the economics associated with the "The Zone," a thirty-mile radius centered on Beverly and La Cienega Boulevards in Los Angeles, outside of which labor union rules make it prohibitively expensive to film (LUKINBEAL 2004). In the final analysis, it does not matter whether a show was filmed in OC or LA; economic concerns dictate that as long as landscape signifies a real place that can pass as the original, realism has been achieved (LUKINBEAL 2006).

LAGUNA BEACH

Like *The OC, Laguna Beach* also depends heavily upon images of the physical landscape to construct place, yet it provides even more spectacular landscapes than those of *The OC*, laying claim to the title of "best-looking show on television" (HEFFERNAN 2005). *Laguna Beach*, different than your run-of-the-mill reality show, has a much higher production value. The show was shot in cinematic style using a widescreen aspect ratio, elaborate lighting configurations, and several special digital cameras, often at sunset, to effectively emphasize the natural beauty of the city and its residents (DEMPSEY 2004; HEFFERNAN 2005).

Establishing shots produce the stunning Laguna Beach setting. In a manner similar to the opening credits of *The OC*, we see, in rapid montage, waves breaking on the beach, Laguna's picturesque coastline, palm trees, a mansion, a tide pool, and several sunsets over water (Figure 42). Again, the camera's gaze does not linger long on these spectacular images; rather our attention is drawn through the credits to the start of the episode. The montage sequence packages landscape as spectacle, which attracts our attention and constructs the place that is televisual Laguna Beach.

Figure 42. Laguna Beach *opening sequence*

The opening sequence, furthermore, is set to teen pop star Hilary Duff's "Come Clean," a song whose lyrics begin with the word "perfect," and continue with the phrase "Let the rain fall down; And wake my dreams; Let it wash away; My sanity," fitting words to accompany such stunning vistas. Paired with this musical cue, metaphor and spectacle merge to produce a "perfect" lifestyle in a "perfect" place, evoking the carefree years of high school, or as expressed by BELLAFANTE (2007), the high school experience we dream we might have had.

Each episode commences with a "Previously on *Laguna Beach*" screen, in which this text is transposed onto a landscape image used repetitively in the series – a landscape that has now become a Laguna Beach icon (Figure 43). Following this establishing shot and the opening credits, the episode title is re-inscribed within a similar repetitive, iconic, spectacular backdrop (Figure 44). Also, shots of cast members' homes and restaurants and shops they frequent become familiar landmarks, reinforcing a *Laguna* sense of place while assisting viewers' navigation of this mediascape (Figure 45).

Figure 43. The iconic Laguna Beach *landscape*

Figure 44. Episode title screen

Figure 45. Transitional shot of cast member's home

Establishing shots, opening sequences, and transitional shots in *Laguna Beach* are emotionally charged and affect our voyeuristic desires, anticipating viewers' interpretation of narrative significance. It is the transitional shots, however, that most effectively establish place in *Laguna Beach.* These shots are often aerial views of Laguna's townscape, Laguna as seen from the ocean, closer views of downtown, or other city icons (Figure 46). Several such landscapes (always different) are presented between scenes throughout the course of each twenty-minute episode to musical accompaniment. Music and landscape move together across scenes, creating affective and emotional currents in animating place and place meanings.

Figure 46. Laguna Beach *transitional shots*

THE REAL HOUSEWIVES

The Real Housewives, a lower-budget cable reality program, also uses spectacle to introduce viewers to life in an Orange County gated community. Rather than the breathtaking beaches and seascapes of *The OC* and *Laguna Beach*, *Real Housewives* features images of inland wealth: expansive homes, expensive cars, emerald golf courses, picturesque palms, bejeweled women, wine, Jacuzzis and swimming pools, horseback riding at sunset. Voice-over narration informs:

> Life is different in a gated community. Land here is one million an acre, the average-priced house is one million eight to two-two. Once people come through this gate, there's a sense of being Teflon-ed – nothing sticks to you. Image is everything in my world. When you're not behind the gates, you don't know what you're missing. This isn't just a place to live, it's a lifestyle.

The penultimate image of this sequence shows Coto de Caza from a distance, miles of homes perched atop a hill, buttressed with the caption "7 million families live in gated communities." The final title screen is a somewhat sinister image of gates closing over uncharacteristically dark skies (Figure 47). As discussed above, a pilot episode is burdened with the task of introducing this setting and convincing viewers that residents of Orange County lead compelling lives. These landscapes, especially the symbolism of the gates, position the ladies of Coto in a place which is highly differentiated, socio-economically, from its surroundings. Meanings associated with the gates and other signifiers of social status establish Orange County as a very exclusive entity. The ominous image of closing gates also conveys a sense that although it may look like paradise, life inside has a darker, dramatic quality.

Figure 47. Introduction to Coto de Caza

In each episode, *The Real Housewives* regularly utilizes landscape to familiarize viewers with the ladies' environment, to construct televisual place. In the opening sequence, we are formally introduced to the ladies and see them engaged in a variety of "Orange County" activities, including stepping into a hot tub with a glass of wine, receiving Botox treatments, drinking a martini on horseback, and trying on diamond necklaces (Figure 48). Transitional shots warn viewers that we will be updated on the goings-on of a different cast member and generally consist of cast members' homes and various bars/restaurants where cast members socialize (Figure 49). Later seasons more regularly use spectacular transitional shots of the Coto landscape – its golf courses, fancy cars, and even Newport Harbor – to re-establish the opulence of the setting and to recapture the viewers' attention. Again, these landscapes ensure that, as viewers are transported across and between the social spaces which comprise televisual Orange County, place-based meaning follows.

Figure 48. The Real Housewives *opening sequence*

Figure 49. Transitional shot of cast member's home

Although the housewives operate in a somewhat different social milieu than the teenage stars of *Laguna Beach* and *The OC* (the grownup version), together, these shows construct a singular televisual Orange County; landscape creates a similar place in all three. Spectacular and metaphorical landscapes presented in establishing and transitional shots all position Orange County as a place of wealth, privilege, and great physical beauty. Images of the beach, warm weather leisure activities, and characters' beautiful homes are presented in each show to orient the audience and to reinforce the cultural meanings contained in this televisual place.

Thus far we have elaborated on landscape's role as cultural product in place-making. Through establishing and transitional shots, landscape serves as an object to be gazed upon, a frame from which to experience the TV world, a conveyor of cultural meaning and geographic realism. These landscapes express broad, rather generic meanings about televisual place. We now turn to landscape as cultural process, to the spaces which produce a more intimate and emotional engagement with televisual place.

LANDSCAPE AS SPACE

Landscape, even when presented and read as a cultural product, is always a work in progress. Landscape is continually shaped, transformed, and discovered as we make our way through it. Through our interactions with landscape, place emerges and comes into being (ROSE 2002). When the function of landscape shifts from place to space, discursive space also shifts from mere background to occupy the foreground of the mise-en-scène. This is a scalar shift; one that moves from wide angle, long range, deep focus, or aerial shots to close ups and spaces of dialogue, events, and actions. Here place is turned into a social space which allows the narrative to unfold, an essential function in televisual place-making.

In *The OC*, discourse space often takes the form of the Cohen home or the pool house where Ryan resides. In the pilot alone, the viewer accompanies Ryan and Seth on a sailboat, to a swanky fashion show fundraiser, and to a party at a beach house. Subsequent episodes feature scenes at a debutante ball, a casino night fundraiser, several lavish parties, and other spaces intended to promote an extravagant lifestyle. Other frequently used settings are the characters' private high school, the real estate development offices where Kirsten works, a diner on the pier, and the beach. These settings undoubtedly work to promote the idea of an extravagant Orange County lifestyle, but it is within social space that these scenes *take place*. It is at the high school and at the parties where viewers get to know the characters, are appraised of their dramatic situations, see the emotion in their faces, and hear the words they speak to each other. It is within these social spaces that Orange County becomes meaningful to its audience, and thus becomes a place.

Because *The OC* is a fictional dramatic series, complex storylines take characters on a variety of adventures in new places which are featured in one or more episodes until the storyline is played to a close. For example, in episode fifteen of season two, the main characters get locked in a mall overnight; the mall is not a regular social space in the series and is only used once as a plot device.

These transient places are not an integral component of the characters' world, but nonetheless expand the mediated Orange County landscape. These places offer new opportunities for social space to unfold; they are literally "meeting places" (cf. MASSEY 1994, 154).

Barred from filming in the local high school, and largely devoid of authority figures, the social spaces of *Laguna Beach* present a very specific slice of teenage life in the "real" Orange County. Through producers' choice of filming locations, Orange County becomes a charmed place where moneyed, photogenic teens can party and surf with worries of neither parents nor school. Approximately forty percent of the show's scenes (from fifteen episodes sampled) are filmed in cast members' homes: a few friends chat while lounging on the deck or laying out by the pool. The home also frequently serves as a party venue, and parents and siblings rarely appear in these scenes.

Approximately twenty percent of the series' social spaces are cafes or nice restaurants (most of the restaurant scenes document ersatz dates). Another twenty percent of scenes are shot at the beach, which proves an appropriate gathering place for gossiping, showing off surfing skills, as well as for breaking up with boyfriends or girlfriends. Viewers also frequently accompany cast members to local boutiques and the nail salon (Figure 50). Occasionally we glimpse these teens at their workplaces, but these are boutiques and surf shops, not burger joints, and therefore function as a convenient place for friends to stop by to chat about upcoming social events. Many other once- or twice-used locations depart from the home-beach-restaurant trifecta, but nevertheless succeed in furthering the show's presentation of a specific Laguna Beach lifestyle. These include a journey to Catalina Island by yacht, a resort in Cabo San Lucas (Mexico), senior prom, a BMW dealership, and the occasional fashion show benefit (note that *The OC* first used this device to depict an Orange County lifestyle in its pilot).

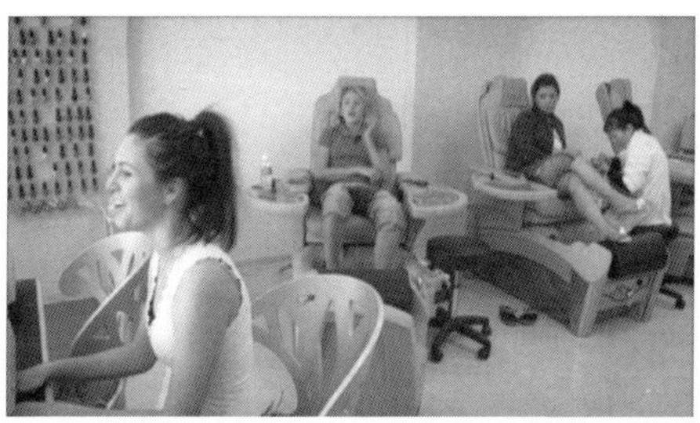

Figure 50. At the nail salon, a favorite Laguna Beach *pastime*

Again, these discursive spaces are not devoid of meaning. Instead, they provide the sites central to place-making. On the deck or at the beach, characters' lives are revealed and these places are made meaningful. Cafes, boutiques, restaurants, and the beach comprise a well-defined mediascape, and participants therein turn these locations into place through practice.

Furthermore, *Laguna Beach* takes advantage of the area's significant physical beauty in a unique way, by combining medium shots of the landscape with narrative action, in what could be termed "Laguna clichés:" characters are often captured having cell phone conversations on their decks at dusk, heartfelt talks at the beach or beach overlooks, or at seaside bonfires in the evenings (Figure 51). While social space is created through characters' emotional revelations within these landscapes, landscape also functions as metaphor and spectacle to enhance the narrative. A gaze off into the ocean, for example, after breaking up with a boyfriend or just prior to leaving Laguna's bubble for college suggests a regretful or reflective state of mind (Figure 52). Here, as elsewhere, metaphor and spectacle serves as an intermediary between place and space, animating the place-making process.

Figure 51. Laguna cliché

Figure 52. Landscape as metaphor and spectacle

By interweaving spectacular, transitional landscapes with this impressive roster of social spaces, *Laguna Beach* constructs Laguna Beach as a parent-less, homework-free world of surfing, swimming, shopping, and partying. Unfettered by the burdens of educational responsibility, landscape produces a place of consumption-based leisure, a lifestyle broadcasted to fans throughout the country and the world.

Similar to *Laguna Beach*, place-making on *Housewives* most often occurs in cast members' homes and in nice restaurants. Because landscape as spectacle or metaphor is less often used in this series, filming locations play a heightened role in crafting mediated Orange County, using up place to create a space within which narrative action will unfold.

The home is a significant component of the housewives' social landscape. As it is much easier to gain permission to film inside a private residence than a public area, roughly fifty percent of each show (from fifteen episodes sampled) takes place in cast members' homes, where the women interact with their families and throw a number of backyard barbeques. Familial relations play a key role in the creation of the homescape, as children and husbands deepen our knowledge of the women's social space and provide drama and conflict. Just as the *Laguna Beach* landscape is all but devoid of parental authority, husbands and other men have a limited presence on *The Real Housewives*; included, it seems, when needed for dramatic purposes. Children are usually featured as spoiled and demanding. These family dynamics produce a social space centered on youth and women.

Nearly thirty percent of filming locations outside the home are bars and restaurants where the women *always* order drinks (Figure 53). Approximately ten percent are trips to salons and spas, and shopping accounts for another ten percent of outside activities. There are also trips to the beach, the Coto de Caza Golf and Racquet Club, and horse stables. Other notable gathering places include an after-work visit to "Vicki's Backyard Spa," hula lessons with Kimberly and her teenage daughter, dinner and a night spent at the swanky St. Regis Resort, a "Botox party" in the office of a Newport Beach plastic surgeon, and a trip back to Illinois for Vicki's twenty-fifth high school reunion. Also, several of the women take trips to Mexico, one goes to Palm Springs for a "weekend with the girls," a family trip is taken to Lake Havasu, and all of the ladies spend a weekend together at a posh hotel in San Diego (which they reach by limo). For the most part, the ladies engage in expensive, enjoyable activities which reinforce a selective Orange County social landscape and maintain the spectacle of a luxurious landscape.

Figure 53. Drinking at a bar/restaurant

The Real Housewives shows life in an Orange County gated community as a montage of eating out and otherwise socializing; rarely does this lifestyle entail an eight-plus-hour workday and grocery shopping. Although we occasionally glimpse realtor Jeana showing a (very expensive) house or see type-A Vicki bustling in her office, the workaday aspects of this "real" life are disproportionately reduced to enhance social places critical to constructing a particular leisure-based OC lifestyle. As does Vicki to maintain her extravagant lifestyle, landscape as social space works hard in this series to create a specific type of place. These spaces draw viewers deeper into the milieu of a Coto resident. At these restaurants, we are privy to the ladies' complaints of their husbands, their kids, and each other, and on these shopping and spa trips, we see them reap the benefits of their wealth, living that nebulous Orange County lifestyle (cf. ARELLANO 2008). Social space is continually transformed into meaningful place as the housewives animate landscape.

LANDSCAPE TENSION MAKES PLACE

In television, landscape functions as both cultural product and cultural process to create place through a meaningful process of becoming. At the start of each Orange County show, viewers are presented with spectacular landscapes in establishing shots. Their function is to frame televisual place, to generate a sense of place through geographic realism, and to imbue it with cultural meaning. The narrative, however, unfolds within discourse space. From broadly construed symbolic landscapes, we zoom into the more personal social landscapes of the characters. Viewers accompany characters as place is enacted within a series of specific settings. As we see this lifestyle played out within the landscape, its social spaces become meaningful and Orange County emerges as a place to viewers. Landscape as space creates place through this process. Viewers are continually reminded of the existence and meaning of this place through repeated presentation of spectacular landscapes in transitional shots throughout each episode.

Thus, central to place-making in television is the ongoing tension between landscape as space and place, as cultural product and process. The second episode of *The OC* provides an illustration. The episode begins with a broad presentation of Orange County as a lucrative setting through a series of ocean and beach shots. These give way to a stunning aerial view of coastal mansions, and next the camera has zoomed in on the Cohen's house. Finally, we cut to Seth and Ryan lounging in the pool (Figure 54). At the pool, the episode's narrative commences, concluding this classic progression from "the general to the particular" (HIGSON 1984: 7). As HIGSON (1984: 7) explains,

> the [scalar] movement from one image to the next rapidly impregnates the space, after each transition, with increasing narrative significance and dramatic purpose, until finally the individual is placed in his environment, as a figure in the landscape.

Figure 54. Place to space in The OC

This progression is "a classical movement from the general to the particular" (HIGSON 1984, 7), and it works on two levels. First, this movement embodies the tension within landscape as a cultural product and as a process. As a product, we are presented with a general way of seeing. As a cultural process, landscape unfolds through the ongoing assemblages of people, places, experiences, actions, and dialogue. Secondly, the movement from the general to the specific is a compositional technique accomplished through "the establishment of fixed patterns of clarity for the variation of scale of shots in a scene" (HEATH 1981, 41). The scalar transition "is of extreme importance attached to providing an overall view, literally the 'master shot' [establishing shot] that will allow the scene to be dominated in the course of its reconstitution narratively as dramatic unity" (HEATH 1981, 41).

After a brief set of scenes alternating between the social spaces of the Cohen home and pool, the opening sequence re-establishes Orange County through spectacular landscapes as a beautiful, privileged place (Figure 39). After a commercial break, we re-enter Orange County and are re-oriented with a shot of the pool house where Ryan resides. The next shot is of Ryan tossing and turning in his bed. The narrative continues as viewers are drawn into the social space of Ryan's current dilemma. At other points in the episode, transitional shots of the Cohen home and other locations reinforce our sense of place before action resumes. The progression and assemblage of different landscape functions (from the general to the specific, from description to narration, from place to space and back again) in each episode and throughout the course of these series constructs televisual place.

The process of place-making as experienced by television viewers is articulated by ROSE's (2006, 547) definition of landscape as a dream of presence:

> Conceptualizing the cultural landscape as a dream of presence means understanding it as an unfolding plane of sensory, affective, or perceptual markers registering and, thus, effecting the emergence of subjectivity.

Televisual landscapes of Orange County enable place to unfold through viewers' affective response to the lives played out within. Although we will never find ourselves truly within the televisual world of Ryan and Seth or the housewives (an "impossible possibilit[y]" [ROSE 2006, 545]), we can imagine our presence in Orange County as we engage with this setting and feel it becoming closer and more familiar – that is, becoming a place – with each episode watched. Viewers use the televisual landscape to "imagine, cultivate, and move towards [the TV] world ... as present and, in the process, to experience it more intensively" (ROSE 2006, 549).

THE MEDIUM OF TELEVISION

It is apparent that the medium of television has much to offer the study of place-making, although the portrayal of place in cinema has garnered much more attention within media geography for several reasons. First, as outlined in chapter one, the TV screen is smaller than a movie screen and, in older models, squarer and unable to accommodate panoramic shots that emphasize landscape. Whereas movies tend to be viewed in total darkness with few distractions, TV programs are fragmented, interrupted by commercials, and are often not the sole focus of the viewer's attention (HIGSON 1987; ADAMS 1992). Additionally, TV shows are shorter in duration than movies, with only 20–40 minutes (after commercials) in which to tell their story. Due to time constraints, constraints of the viewing frame, and lower production budgets, geographic realism is not always a priority in TV. Television generally emphasizes action occurring within social space over the broader landscape meanings (landscape as place, spectacle, and metaphor) that cinema more often employs. Development of place on TV most often occurs speedily and thriftily, with the use of only a few establishing shots. These characteristics of television viewing are assumed to reduce television's ability to utilize landscape in the construction of place.

This trio of shows does much to negate these arguments. *Laguna Beach*, a ground-breaking show in many ways, disproves the notion that televisual landscapes are necessarily less stunning than those on the big screen. *Laguna Beach*, like cinema, is shot with a widescreen aspect ratio and elaborate lighting configurations, so that "blonde hair and tanned skin emit an especially painterly glow" (MCCARTHY 2004), or so that "every frame ... looks like it has been dipped in honey" (ROCHLIN 2005). The growing popularity of widescreen high-definition TV sets will undoubtedly encourage a trend toward increased use of panoramic landscapes on television. Additionally, reality TV has become popular with networks because it is economical (non-actors, no sets). Thus, in this genre, more resources may become available to interweave the broader place-based meanings expressed by spectacular or metaphorical landscape with the micro-level social spaces already present in reality programming.

In all three shows, Orange County holds the title role; their premises depend upon scripting evocative meaning onto this place. During a period when many shows, in the interest of time, have dispensed with title sequences and theme songs (PEYSER 2005), or reduced them to a logo shown for a few seconds with a few notes of music, *The OC* and *Laguna Beach* use their opening credits to introduce place to great effect. Unlike in the movies, these sequences are seen week after week and sense of place is reinforced through repetition with each viewing. Time accrued through the course of a season allows landscape to build and maintain televisual place.

The strength of television's place-making power thus lies in its episodic and repetitive structure. Fans visit the TV world on a weekly basis, often over a period of years. As described by BRUNO (2007, 39–40), fans are nourished by the familiarity of this place and its landscapes, and these satisfy the desire for voyeuristic travel:

> Cultures and individuals fixate on specific landscapes for different reasons and reactively pursue them. A traveler seeking a particular landscape may go there, even filmically, to be replenished, restored, held, and fed. In the hub of traveling and dwelling, we are absorbed in the stream of emotions and experience an embracing affective transport.

The OC aired 104 episodes over its four-year run; *Laguna Beach* (including its short-lived spin-off, *Newport Harbor)* ran fifty-eight; and, *The Real Housewives* eighty-two episodes in six seasons (and counting). The hours, weeks, and years encompassed by the span of a series allow significantly *more* time for landscape to do its work and for place-making to occur than does a movie viewed only once. Landscapes of television play to our topophilic and scopophilic desires, allowing the places they create to become an integral part of our daily lived experience (LUKINBEAL 2005). Televisual place-making is, therefore, a dynamic and ongoing process which unfolds as we watch and wait to return.

In this chapter, we have described the ways in which landscape contributes to the place-making process and works to create televisual Orange County. The next chapter explores the connection viewers forge to place within the social spaces of television. To do so, we address the emotional component of televisual place as communicated through the genres of melodrama and reality TV.

CHAPTER 5

EMOTIONAL REALISM

As we have seen in chapter four, landscape provides the milieu within which televisual place unfolds. It is via these homes, restaurants, and spas that we forge a connection to the characters, their lives, and the places they inhabit. This chapter further addresses the place-making process as it occurs within narrative space by exploring the concept of emotional realism within the primetime soap opera and dramatic reality television. After establishing the role of emotional realism in televisual place-making, we conclude by offering a means of conceptualizing the televisual experience.

EMOTIONAL REALISM AND THE PRIMETIME SOAP OPERA

The OC is best described as a primetime soap, among the latest in a genre that the massively popular Reagan-era sagas *Dallas* and *Dynasty* introduced. In the 1990s, the genre's subject shifted from wealthy patriarchal families to wealthy teens and young adults with *Beverly Hills 90210* and *Melrose Place* (FEUER 1995). As evidenced by this lineup, U.S. soaps tend to attract viewers with their depiction of an aspirational lifestyle in congruence with the American Dream (JHALLY and LEWIS 1992). As such, these popular programs share an opulence of setting, an excess in the melodramatic tradition (FEUER 1995): a fancy ranch in oil-rich Dallas, the lavish mansions and automobiles enjoyed by *Dynasty*'s ruling family, the trendy halls of West Beverly High, and of course, Orange County's yachts and oceanfront property.

Primetime soaps differ from daytime soaps in their weekly broadcast, higher production values, and increased attention to visual excess, but are similar in their melodramatic nature (ANG 1985). The soap opera, or serial melodrama, operates upon a "heightened emotional register" and can be identified by several characteristics (JOYRICH 1996, 51). First, these programs feature several storylines across a "community" of characters. Second, they focus on the private, personal lives of these characters, leaving much of their participation in the public sphere untold (ANG 1985). Finally, the characters of melodrama are trapped aboard a never-ending emotional roller coaster; they are continuously bombarded with devastating news, emergencies, and tumultuous relationships (ANG 1985; JOYRICH 1996). Until the series finale, there is no conclusion and nary a happily ever after.

The results of several reception studies suggest that much of the pleasure derived from watching television stems from a degree of realism present in its programming (cf. ANG 1985; JHALLY and LEWIS 1992). If a show contains something that viewers can believe in, or relate to, they are more likely to suspend their disbelief, abandon their doubts, and become more thoroughly immersed in the televisual experience. As discussed in the previous chapter, landscape constructs a setting within which the drama takes place (LUKINBEAL 2006). But this is only one part of the equation. For a program to be truly enjoyable, the viewer must also identify with the characters and their situations. In the words of ANG (1985, 20; paraphrasing Bourdieu), "popular pleasure is first and foremost a pleasure of recognition."

Their intricate storylines, unrelenting drama, and alarming number of amorous relationships within the character community make soaps seem, at first blush, quite unrealistic and campy; they are often derided in terms of their far-fetched theatrics. Accordingly, ANG's (1985) study of *Dallas* questions the immense popularity of this show. Through viewer responses to the question of why they enjoy watching *Dallas*, ANG (1985, 45) finds that the appeal of the series, and melodrama in general, lies in what she terms "emotional realism." Many elements of melodrama are unrealistic on a denotative, or literal, level but are recognizable at a connotative level as associated meanings. Rather than by an empirical realism measuring a text's accurate reproduction of reality, fans judge melodrama by its ability to evoke emotions. Often in television, states ANG (1985, 45), "… what is recognized as real is not knowledge of the world, but a subjective experience of the world …"

The emotional realism of melodrama is built upon a "tragic structure of feeling" (ANG 1985, 46) or a "complex of meanings" (ANG 1985, 61) which evoke the never-ending search for happiness that characterizes life. Melodrama exaggerates these highs and lows for emotional effect. Characters endure countless traumas as if such frequency were normal, and viewers read these as metaphors for their own "life's torments" (ANG 1985, 63). Alcoholism, for example, is an oft-used plot device in both *Dallas* and *The OC*. In such shows, heavy drinking is not so much addressed concretely, as the disease it is, but functions to symbolize a character's suffering and fragile emotional state. Regular viewers of this genre readily interpret the metaphor and lend it emotional significance (ANG 1985). Fan responses illustrate *Dallas*' emotional impact:

> *The reason I like watching it is that it's nice to get dizzy on their problems … I really enjoy having a good cry with them. And why not? In this way my other bottled-up emotions find an outlet* (ANG 1985, 49).

> *It is easier and loads more pleasurable to dash away a tear because Ma Dallas has cancer – because she's only acting – than because of whatever annoying things are in your own life, and at least you have got rid of that tear* (ANG 1985, 50).

ANG (1985, 79) concludes her analysis by explaining how possession of (and openness to) a "melodramatic imagination" is key to taking pleasure in soap opera's emotional realism and its articulation of the feelings we don't often have the means or opportunity to express. This is why some like to watch *Dallas* and other shows of its ilk:

> The melodramatic imagination should be regarded as a psychological strategy to overcome the material meaninglessness of everyday existence, in which routine and habit prevail ... In other words it is a matter of making the "world we inhabit one charged with meaning" ... The melodramatic imagination is therefore the expression of a refusal, or inability, to accept insignificant everyday life as banal and meaningless, and is born of a vague, inarticulate dissatisfaction with existence here and now. This then is the tragic structure of feeling: it is not about the great suffering which plays such a prominent role in the history of humankind ... but is rather about what is usually not acknowledged as tragic at all and for that very reason is so difficult to communicate. There are no words for the ordinary pain of living of ordinary people ... for the vague sense of loss ... By making that ordinariness something special and meaningful in the imagination, that sense of loss can –at least for a time – be removed. It is in this world of the imagination that watching melodramatic soap operas can be pleasurable: *Dallas* offers a starting point for the melodramatic imagination, nourishes it, makes it concrete (ANG 1985, 79–80).

According to ANG (1985, 82), *Dallas* penetrates viewers' "imaginative world" and "connects up with one of the ways in which [viewers] encounter life." Specifically:

> In so far as the imagination is an essential component of our psychological world, the pleasure of Dallas ... is not a *compensation* for the presumed drabness of daily life, nor a *flight* from it, but a *dimension* of it ... a life without imagination does not exist (ANG 1985, 83) (Figure 55).

Figure 55. Pamela (Victoria Principal), following a spat with husband Bobby on Dallas

In ANG's (1985) view, the emotionally-elevated world of TV melodrama grips our imagination and in so doing, becomes an integral part of everyday experience. We long for feelings of emotional affect and willingly "play along," actively appropriating personal meaning from TV's mass message. Emotional realism extends our lived world to encompass these televisual places, allowing us to join a new community and to participate in its place-making. Place is created as these emotions are emplaced and, with time, built up within the social spaces of a television series. Melodrama is designed to draw viewers into its world, and its world into our lives.

Yet, emotional realism works on two levels: it concurrently allows participation in place-making, while also offering disengagement – a way out – from the tricky situations implicated in this process. Emotional realism simultaneously connects viewers to televisual place while also sheltering viewers from the effects of emotional entanglement. We could hardly survive a life so filled with drama, yet characters' constant upheavals pleasurably expand our affective experience. We are simultaneously insiders and outsiders in the televisual world. We emotionally engage with televisual place, yet these places remain "dreams of presence" (ROSE 2002, 547) within the melodramatic imagination. We may experience the emotional climax of *The OC*'s season one finale (see synopsis in chapter three), or "dash away a tear" with "Ma Dallas" (ANG 1985, 50) but, when we are ready, we can also forget about Ma Dallas' troubles and return to our quotidian existence. This is, undoubtedly, an alluring aspect of participation in televisual place.

REALITY TV

Reality TV exploded onto American airwaves in the new millennium. The breadth of shows under the reality umbrella include *Bachelor*-style dating shows, to *Survivor* and *American Idol*-type competitions, to voyeuristic "episodic reality soaps" which offer a window onto the lives of celebrities or other people of interest; *The Real Housewives* and *Laguna Beach* fall under the latter category (HOLMES and JERMYN 2004, 6). All reality shows share characteristics of minimal writing and the casting of, if not exactly "regular folks," at least non-actors. News and documentaries are excluded from this genre because in reality TV, "access to the real is presented in the name of dramatic uncertainty, voyeurism, and popular pleasure," rather than public service, education, or more altruistic goals (OUELLETTE and MURRAY 2004, 2). Though in name the genre implies a more realistic portrayal of the people and situations it depicts than overtly fictional sitcoms and soaps, much evidence points to the contrary. The "real" in reality TV is instead an "entertaining real" (OUELLETTE and MURRAY 2004, 4) or a "popular reality" (HARTLEY 1996, as cited in CORNER 2002, 268). As FRIEDMAN (2002, 8) explains, "What is new and fresh about these programs is not their allegiance to reality but the marriage of reality conventions with dramatic structure."

In fact, the appeal of reality entertainment owes much to the conventions of melodrama. The "real" people cast in these shows are placed in contrived situations created for maximum conflict and suspense: either trapped on an island to compete for one million dollars (*Survivor*) (Figure 56), picked to live in a house with seven strangers (*The Real World*), or entrenched in competition for the affections of the perfect guy (*The Bachelor*). MCCARTHY (2004, 1) observes: "reality TV remains firmly within the realm of melodrama, dependent for its appeal on the ability of characters to externalize emotions and internal conflicts through speech, expressions, and gestures."

Figure 56. Recent Survivor *cast members*

Although the situations in episodic reality soaps are not so obviously contrived as those of "gamedocs" like *Survivor*, much of their drama is in fact created in the editing room. MTV's successful series *The Osbournes*, "the first reality sitcom," as KOMPARE (2004, 107) points out, chronicles the family life of aging rocker Ozzy Osbourne in a manner drawing heavily from the conventions of traditional fictional sitcoms (Figure 57). Hours of footage were edited into cohesive twenty-minute episodes within which "problems are introduced, conflicts play out, and the end of the episode reaches a mild resolution" (KOMPARE 2004, 108).

Figure 57. The Osbournes *(2002)*

In the case of reality soaps like *The Real Housewives* and *Laguna Beach*, subjects are "cast and planned as carefully as in any fiction" (KOMPARE 2004, 106).

Reality producers search for "telegenic" people who look and are prone to act in ways which fall within the visually pleasing and dramatic norms of the medium (i.e. people who are both good-looking *and* emotional) (KOMPARE 2004, 106). Also, just as *The Osbournes* accomplished in sitcom style, editors of these shows advantageously use known conclusions to construct episodes containing the pleasurable codes of melodrama (FRIEDMAN 2002). In *Housewives*, for example, some storylines are contained within one episode, such as Jeana buying her daughter a new car, and others build toward a crescendo. Episodes often end on an uncertain note, and as promised in previews, we will continue to follow this issue in the next episode. To illustrate, the big story in the first season of *Housewives* is the tenuous nature of housewife Jo's relationship with her fiancé, Slade, and the season ends with their break up (Figure 58). Previous episodes lead to that climax by emphasizing Jo's boredom and desire to establish her own career, as well as Jo's proclivity to "go out with the girls." A local critic comments:

> I think we all know that Scott Dunlop [*Housewives* creator] could never have sold a show about his hometown had he focused on the mundane lives these women lead much of the time (MICKADEIT 2006c).

Figure 58. A tense moment in Jo and Slade's relationship

The Real Housewives is a staid member of the reality genre, and it thus utilizes another convention of reality TV: the use of videotaped interviews which provide a character's description and explanation of the situation after-the-fact. Because there is no script, these interviews fill in narrative gaps that the footage cannot be edited to express (MCCARTHY 2004). Emotional realism on reality TV is thus composed in this manner: Cast members emplace their drama by acting within social space; they later explain their feelings to the camera; editing molds material into episodes conforming to television norms that viewers readily understand and are primed to identify with; and ultimately, viewers become ensconced in Orange County. Each episode builds upon this formula to construct televisual place.

Laguna Beach, proclaimed the first "reality drama," differs considerably from shows like *The Real Housewives* (CHANG 2004), not only in its cinematic look.

Although the text contained in its opening screen (and subtitle: *The Real Orange County*) firmly presents the show as within the reality genre (Figure 59), *Laguna Beach* much more resembles traditional melodrama in structure. It does not utilize the interview-narrative format and, as MCCARTHY (2004) further observes:

> … many scenarios are clearly staged for the camera. We see teenage boys squirm and mumble as they endeavor to carry on a group conversation on a set topic: will they stay in touch after graduation? And we witness both ends of telephone conversations, a strategy that signals the show's commitment to narrative form and continuity over the pretense of spontaneous action. These staged moments position the teenage cast as improvising actors rather than sociological subjects. Together with the show's lush cinematography, they forge a connection between *Laguna Beach*'s "real Orange County" and the dramatic show it aims to supplement: Fox's lavishly shot teen soap hit *The OC* (Figure 60).

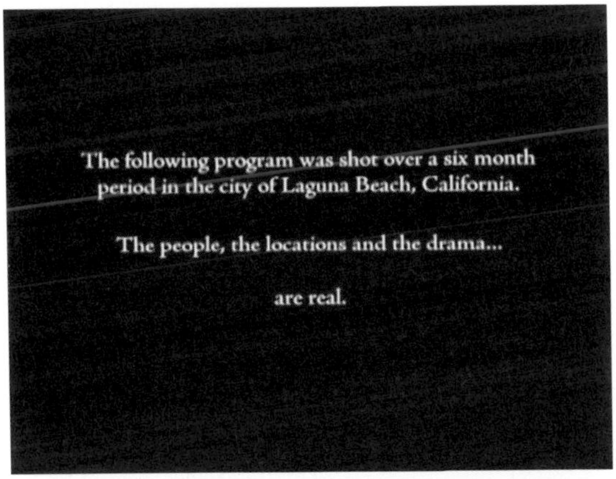

Figure 59. Laguna Beach*'s claim to the real*

Figure 60. Typical Laguna: discussing the evening's plans against a stunning backdrop

Several devices are used to promote emotional realism in the melodramatic genre: a reliance upon emotive music, in lieu of speech, to express characters' deepest feelings; close-up shots linger a few seconds after the last line of dialogue in a scene is uttered in order to emphasize the emotion contained in the actor's facial expression; and rather than attempting to portray the nuanced complexity of many "real-life" personalities and situations, characters and the decisions they make are painted in easy-to-read oppositions of good versus evil and right versus wrong (FEUER 1995; JOYRICH 1996).

These generic devices are put to work within the social landscapes of *Laguna Beach*. *Laguna Beach* viewers participate in an emotionally-charged place-making process more similar to that of *The OC* than *The Real Housewives*. Frequent use of musical montage tells a good deal of the story in each episode (MTV's *Laguna Beach* mix is available for purchase on iTunes), as does the "liberal inclusion of long close-ups of the characters' relatively blank faces" (LEVINE 2006) (Figure 61). LEVINE (2006) additionally notes that, as in daytime soaps, nothing really happens in any given episode of *Laguna Beach*. "Instead, each episode revolves around 'drama' (the kids' favorite word …) which has the vague meaning of some kind of conflict or emotional upheaval," usually involving the show's infamous "not-entirely-there" love triangle (LEVINE 2006). The many conversations we witness at restaurants and the beach "work to force a plot around a sequence of somewhat disconnected events and characters" (LEVINE 2006) (Figure 62).

Figure 61. Lauren's thoughts drift from the bonfire

Figure 62. Jason and Jessica conclude their tumultuous relationship in season two

As an astute commenter on LEVINE's (2006) (online) article observes, however, the "thinness" of the characters and lack of strong narrative content "imitates the banality of real life." In a way, the emotional realism of this "reality drama" is heightened by the normalcy of events it depicts – chatting with friends, dating, getting pedicures, going to prom. If, as ANG (1985) argues, primetime soaps draw their appeal from identification with characters' emotional reactions to over-the-top situations, then *Laguna Beach*, in its "meaning without meaning" (LEVINE 2006) manages to imbue less meaningful situations with a similar emotionality. Orange County is associated with a powerful affective response, a result impossible without music. LUKINBEAL and ZIMMERMANN (2008, 21) explain:

> Audio guides our affective response. It saturates film and … gives space depth through emotive becoming, and guides the unconscious experiential musings of the mind. It calls on our subjectivity to react to seen, heard, felt, and experienced memories and emotions. The assemblage of audio-video takes us on a voyage through ourselves as much as through the film's narrative.

Therefore, reality TV does not mean more realistic narratives or more realism; its contrived situations, carefully constructed episodes, and *Laguna*'s music-based affect are evidence that reality TV draws its appeal not so much from a more accurate portrayal of everyday life, but rather from devotion to melodramatic conventions linked to emotional realism. In short, *Laguna Beach,* as compared to *The OC*, is more akin to high school as imagined (paraphrasing BELLAFANTE 2007). In contrast to the excessive world of fictional soaps, the banality of reality programming offers a means of place-making writ small, which occurs through a focus on interpersonal relations, built over time, within a familiar social landscape (Figure 63).

Figure 63. The senior girls toast high school memories

Both the primetime and reality soap are designed to resonate with viewers on an emotional level, and this emotional realism is integral to the place-making process. Landscape contributes to this process by establishing geographic realism, which allows the social spaces of the narrative to "take place." Landscape links cultural meanings, stereotypes, and icons to emotional realism. Emotional realism, as conveyed through conventions of melodrama within the social landscape, generates a powerful affective response, emplaces these sensations, and expands the reach of our lived experience. This engagement creates "the OC" as a meaningful place, and this connection builds and strengthens with each episode watched, each hour spent there. The experience of watching TV involves far more than just seeing, for film and television are not simply images on a screen, but "an assemblage of sight and sound, of texture and (e)motion, memory and experience" (LUKINBEAL and ZIMMERMANN 2008, 15).

THE REALITY OF TV

Despite the claims of reality TV to represent life (in place) as it actually unfolds, viewers are well aware that much of its content is made-for-TV (HILL 2002; OUELLETTE and MURRAY 2004). Both primetime and reality soaps are fabricated forms, and so rely upon similar expressions of realism – geographic realism to convince viewers to suspend their disbelief and to construct the spaces and places of narrative action, and emotional realism to thoroughly engage viewers within, making televisual place "matter." Thus, the paradox of television is that "we know that these characters are not real, yet we gain pleasure from them in part because they *seem* real" (JHALLY and LEWIS 1992, 22). We are fully aware of television's fiction, but its realism allows it to become a reality that matters and is meaningful in our daily lives.

When asked to judge the realism of *The Cosby Show*, the popular 1980s sitcom, participants in JHALLY and LEWIS' (1992, 28) reception study frequently commented that the show was both "very realistic" and "totally unbelievable" (Figure 64). ANG's (1985, 50) audience survey also indicates "a constant to and fro movement between identification with and distancing from the fictional world ..." JHALLY and LEWIS (1992) go on to posit that lacking enough direct experience upon which to base a judgment or form an opinion, the mediated world fills this gap; we use it to build our own *musée imaginaire*. Citing a study by PRESS (1991, 163) as evidence, when a female respondent is asked to compare a female attorney she once met with Clair Huxtable (matriarch of *The Cosby Show*), she believes in the mediated version:

Interviewer: *You know one woman attorney. Did she remind of you Clair?*
Respondent: *No. Not even closely. Because she wasn't as feminine as Clair Huxtable.*
Interviewer: *Who do you think is more typical of women attorneys?*
Respondent: *Clair Huxtable.*
Interviewer: *Why?*
Respondent: *Well, because I've seen other ones on television like on the news and things like that and they are all more feminine than the one I knew.*

Figure 64. Cast of The Cosby Show *(1984)*

This response is a remarkable example of the power of television to influence perception and the undeniable attraction of television's simplified, stereotyped, and exaggerated depictions of people and place. JHALLY and LEWIS (1992, 30) reach a straightforward, yet important conclusion: "We can no longer assume that the content of TV fiction does not matter simply because TV viewers understand that it is fiction." In short, our judgments and opinions are not always based on a strict distinction between "fact" and "fiction;" more often than not, this line is blurred by the media.

In a subsequent article, LEWIS (2004) expands upon these findings to offer a compelling reconsideration of how we qualify direct versus secondhand experience. He begins by recognizing, as does ANG (1985), that television is an integral component of our everyday experience:

> We cannot systematically pluck those hours of the day we spend watching it and disregard them as nothing more than a vicarious, secondhand class of experience. When we turn off the TV we do not suddenly return to a more vivid [or less vivid in the case of TV images], less ephemeral space (LEWIS 2004, 292).

Instead, just as life is not lived without imagination (ANG 1985), the mediated and the material coexist, overlap, and are routinely juxtaposed: "we eat dinner, watch TV, talk about TV, do the dishes … it is part of us, and we feel that we are part of it" (LEWIS 2004, 295).

Furthermore, as LEWIS (2004) implies above, and as the personal nature of emotional identification with melodrama suggests, watching TV is not a vicarious experience of material place, but a direct experience of mediated place. This view is supported by KIM and RICHARDSON's (2003) reception study of Vienna, Austria through the film *Before Sunrise* (Figure 65). KIM and RICHARDSON (2003) had hypothesized that viewers would experience Vienna vicariously, through empathetic involvement with the film's characters, but instead found that empathy had little to do with the changes in viewers' perception of this place. They conclude that

> more direct (rather than vicarious) participation with a place may occur through movie viewing. That is, subjects may feel that they are personally and more directly drawn into the 'presence' of a place, rather than experiencing it through the eyes of another (KIM and RICHARDSON 2003, 231).

Instead, we experience place through the camera's eye which "has penetrated so deeply into reality" that it presents "an aspect of reality which is free of all equipment" (BENJAMIN 1968, 233–234). The camera's eye is the same as our own; its presence is our presence.

Figure 65. Before Sunrise *(1995)*

In 1973, a documentary-style program called *An American Family* was produced by PBS as a sociological experiment. Often cited as the first reality show, this exercise in cinema verité filmed Santa Barbara's Loud family in their home over a period of several months (OUELLETTE and MURRAY 2004) (Figure 66). The novelty of this program prompted BAUDRILLARD (1994, 28) to describe the paradox of directly experiencing mediated place:

> The producer's triumph was to say: "They lived as if we were not there." An absurd, paradoxical formula – neither true nor false: utopian. The "as if *we* were not there" being equal to "as if *you* were there. It is this utopia, this paradox that fascinated the twenty million viewers, much more than did the "perverse" pleasure of violating someone's privacy. In the "verité" experience it is not a question of secrecy or perversion, but of a sort of frisson of the real, or of an aesthetics of the hyperreal, a frisson of vertiginous and phony exactitude, a frisson of

> simultaneous distancing and magnification, of distortion of scale, of an excessive transparency. The pleasure of an excess of meaning, when the bar of the sign falls below the usual waterline of meaning: the nonsignifier is exalted by the camera angle. There one sees what the real never was (but "as if you were there"), without the distance that gives us perspectival space and depth vision (but "more real than nature").

Figure 66. The Louds of An American Family *(1973)*

In the case of television or film, we experience voyeuristic pleasures when viewing a spectacular establishing shot of the Laguna Beach coastline, as well the haptical pleasures of being present at an intimate scene. Our televisual experience is not limited to the optical plane, but also includes the anti-optical, or experience that is directly felt as we accompany characters on their narrative journey through place. Filmic allure moves between the optical and the haptical; we are granted the pleasures of being the omnipresent voyeur and simultaneously the pleasures of engaging with people, places, and situations through our own experiences, feelings, and emotions. Thus, a television program is the culmination of melodramatic production techniques designed to promote emotional realism and to evoke a sensory reaction – a uniquely televisual experience of place through its screen. The emotions and sensations we feel through our active personal and meaningful engagement produce a gathering place.

Upon visiting the U.S., BAUDRILLARD (1988, 104) proclaimed, "The cinema and TV are America's reality!" Indeed, television expands our experience in and of place, and this experience is a part of our everyday lives. To paraphrase SCHOENKOPF (2008), if perception is reality, then *The Real Housewives* show us the real Orange County. Through the place-making process, television is transformed into real virtuality, not merely an image of the real world, but experience itself. While the experience of physically being in a place and watching that place on TV are epistemologically different, they are ontologically the same. Place is always a way of knowing the world and a meaningful experience in it, and integrally, place-making is always a process: "... a viewer's participation in a film is dynamic and continually changing as the text evolves, and with the work of narrative convention" (AITKEN and ZONN 1994, 20). We live in film, and film lives in us (KENNEDY 2008).

EMOTIONAL AND GEOGRAPHIC REALISM

In this and the previous chapter, we have explored the creation of place on TV. The production techniques involved in creating a sense of both emotional and geographic realism constitute the processes of place-making initiated by the medium (rather than the audience). Realism, of both types, is the key to suspension of disbelief and subsequent audience engagement with the televisual world. As discussed in chapter four, landscape as place and spectacle, as seen in establishing shots, the opening credits, and transitional shots of a television show, serves to script and establish geographic realism, enticing viewers to believe in this place and to inscribe it with certain cultural attributes.

The large-scale landscapes which introduce place at the beginning of a show soon fall away to smaller-scale narrative space, within which emotional realism becomes the dominant mode of place-making. Through the camera's eye, we are present within these social spaces and are haptically drawn into place through an emotional resonance engendered by hearing intimate conversations and seeing expressions in close-up, by participating, emotionally, in the televisual world.

After a commercial break, or between scenes, we again find ourselves in a "position of visual mastery" (HIGSON 1984, 19), gazing upon spectacular landscape in a transitional shot. After a moment, we are brought back into narrative space and into the presence of place. We are no longer gazing upon televisual place, but actively experiencing it. Described in chapter four as a progression between landscape as space and place and back again, the continuous scalar movement between the conventions of geographic and emotional realism is what makes televisual place. Broad meanings contained in the large-scale landscapes that express geographic realism ground the personal meaningfulness engendered by emotional realism, as felt within the small-scale social spaces of the television world. The constant ebb and flow of geographic and emotional realism, cultural meaning and personal meaningfulness, or in Orange County, the interchange of shots of the beach with interpersonal scenes in homes, restaurants, and hot tubs, compose and comprise televisual place.

This chapter has shown that televisual place is not limited to the visuality of its landscapes, but also includes the emotional, haptical, and affective response of our lived world. Place-making, even through television, is a panoply of direct experiences. This and the previous chapter have demonstrated the processes by which the production techniques of geographic and emotional realism create televisual place, but place-making, in the end, is dependent upon viewer engagement. The next chapter engages the processes of televisual place-making from an audience perspective.

CHAPTER 6

THE REAL ORANGE COUNTY

The experience of televisual place is not just vicarious, voyeuristic, or indirect. Viewers affectively engage in televisual place-making. Melodrama's emotional realism works to connect viewers with the show, and this connection evokes a personal journey through the televisual world and one's own subjectivity. This chapter uses audience survey information to explore ways in which viewers experience OC television shows and engage in televisual place-making.

SURVEY METHODOLOGY

In order to uncover the experience of televisual Orange County by its audiences, we surveyed 169 college students in the American Midwest during June and August 2008. The surveys consisted of a combination of multiple choice and open-ended questions about their travel experience, TV viewing habits, and knowledge and impressions of Orange County as gleaned from the television shows included in this study (Appendix A). The students surveyed attended the University of Nebraska-Lincoln, Iowa State University, Des Moines Area Community College, and the University of Iowa, and were enrolled in summer courses in a variety of disciplines. We recruited these participants by contacting summer course instructors and asking permission to administer the survey during class. The students who chose to participate were overwhelmingly from the Midwest, mostly Iowa and Nebraska, with few exceptions. Only twenty-five respondents (15%) had visited Orange County. Sixty-three percent had seen either *The OC, Laguna Beach*, or *The Real Housewives*; but only twenty-nine (17%) identified themselves as fans. We chose to target students from the Midwest based on the assumption that, because they live far from Orange County, they knew little about it prior to watching these shows. Lacking previous knowledge, television is more likely to affect their perceptions of Orange County. Also, as indicated by a producer of *Laguna Beach*, these shows were created to appeal to people outside California and its not-so-sunny, beach-less environment (SCHEIBAL 2008).

In October and November 2008, we conducted a similar online survey specifically targeting fans of *The OC, Laguna Beach*, or *The Real Housewives* (Appendix B). Participants in the online survey were recruited using the "snowball" method. This method entails asking an initial set of participants to identify additional participants (BRADSHAW and STRATFORD 2000). We emailed a request for participation to friends, Arizona State University graduate students, and other email contacts, and asked that they forward the link to others who are fans of the shows. We also posted a link to the survey on fan webpages and discussion boards related to the shows. Ninety-six people completed this survey, ranging in age from fourteen to fifty-six, with a median age of twenty-six. The participants were well-distributed throughout the country, from California to Philadelphia, with a few even from outside the U.S., including England, Canada, Russia, Spain, India, Lebanon, and South Africa. We chose to target fans on the assumption that people who actually follow and enjoy the shows (and identify themselves as fans) have more to say regarding their experience than do casual viewers and thus more actively engage with televisual place.

In November 2008, we conducted a survey of 165 students at Saddleback Community College and Orange Coast Community College in Orange County (Appendix C). These students hail from all parts of Orange County, some are from or near the cities in which the shows are set, and some are from very different parts of the county. Thirty-eight (22%) are or were fans of at least one of the shows. The Orange County students served as a control group to which we asked many of the same questions used in the other surveys. As residents, they are intimately familiar with the area and would therefore be more likely to compare the "real" Orange County with televisual Orange County. Unlike residents of Orange County, the Midwestern and fan survey groups are outsiders with much less experience in the absolute location and therefore may be more inclined to believe in "the OC" and affectively participate in televisual place-making.

Open-ended survey questions posed to all groups generated responses that reveal different ways of encountering mediated place and modes of engaging in televisual place-making. Responses were classified as one of three types: comparative, subjective, or incorporative. Comparative-type responses (and the questions which prompted them) reinforce the naturalized distinction between objective reality and the televisual world. For example, the question "How much of your knowledge about Orange County comes from these shows?" and an answer such as "*I think those shows are fake and aren't what real life is like*" maintain the real/reel binary and position televisual OC as a mere re-presentation of the physically-existing place. Comparative responses highlight the tendency to make sense of televisual place as a fictional world that does not correspond to the reality of a locale. Those who respond in this manner experience television for purposes of entertainment and do not perceive themselves as active participants in its world. This point of view functions primarily at an optical level, in that televisual place is seen and experienced on a voyeuristic plane.

Subjective responses indicate that television has affected viewers' worldview, but a worldview based on stereotypes. These participants exhibit a belief in the televisual world, but maintain the distinction between mediated place and their own lives, or an outsider relationship to televisual Orange County. Subjective responses are similar to comparative ones, but they do reveal some affective impact on subjectivity. They indicate that the respondent has, to a degree, participated in place-making, but it is "thinly" construed via the emplacement of stereotypes and lacking in depth. A subjective response to the question stated above ("How much of your knowledge about Orange County comes from these shows?") is: "*I see the people [who live in Orange County] as rich, rude, and not very mature.*" This type of response moves beyond the "suspension of disbelief" to affectively merge the real/reel into a composite stereotype.

Finally, the survey questions elicited some responses which can be described as incorporative, indicating identification with the show's emotional realism and its work as an affective agent. We use the phrase "affective agent" to relate to what THRIFT (2008) calls affect programs theory. SHAW and WARF (2009, 1338) suggest that this is the "most commonsensical interpretation of affect. This view is derived from a hermeneutic of emotions such as surprise, fear, anger, disgust, contempt, sadness, and joy" (SHAW and WARF 2009, 1338). These types of responses dissolve notions that televisual place isn't real or even that it re-presents reality, demonstrating that it is a place in and of itself. Incorporative responses position televisual place as simulacral, where a reality-effect is created through audience interactions with the televisual world (DOEL and Clark 2007; LUKINBEAL and ZIMMERMANN 2008). Rather than inscribing televisual place with cultural or stereotypical meaning, these respondents fold televisual place into their own lived worlds, by "place[ing] themselves in relation to [its] specific features …, in such a way that their meanings may be revealed or disclosed" (INGOLD 1993, 171). Televisual place-making is never a completed process of incorporation, as televisual place is not unlike one's childhood home, a place to re-visit "to be replenished [and] restored … [to be] absorbed in the stream of emotions and experience an embracing affective transport" (BRUNO 2007, 40).

SURVEY RESULTS

Comparisons between televisual OC and the "real" Orange County are easy to make and difficult to avoid. We are taught not to believe everything seen on TV, that TV is not real or objective, and strict adherence to this point of view relegates participation to voyeurism, an observation of sensational subjects. Most of the survey questions furthered this separation and encouraged respondents to think about televisual place not as a unique composition of production techniques which evoke an emotive journey, but as a re-presentation of the Orange County in which people physically reside. This mode of evaluating mediated place diminishes its substance, and reinforces the normative belief that TV and cinema are re-presentations of some ontologically stable reality. Reality shows, in particular, promote acceptance of this viewpoint and effectively naturalize the belief that TV is a re-presentation of a place and its people (see Figure 22). As argued above, even reality shows are simulacral. While they may depict real locations and people, they do not show "actual" events or "actual" social geographies that one finds in Orange County. Rather, TV creates its own places.

One question from the survey begs such a comparison: "Do you think these shows portray Orange County in a realistic manner?" This question is positioned within the normative belief of re-presentation and influences respondents to answer in kind. Unsurprisingly, a majority of Midwestern respondents answered "*no*." Some elaborated that "*not everything is that perfect*" or "*not everyone there is rich*." Others, however (approximately 20%), usually admitting to having no prior knowledge of Orange County before seeing these TV shows, responded subjectively, indicating that television has influenced their perception of the place it supposedly depicts. For example: "*I've never been there, but seeing these shows creates no desire to go*" or "*I think most people there are living well, or at least better than people in Nebraska*." Also as expected, Orange County residents responded similarly to the same question, expressing a distanced relationship with televisual Orange County, one that is very different from their own lives in the same location. Generally, "*not everyone who lives in Orange County is rich*." Comparisons of the "real" and the "reel," and the questions which presuppose such a distinction, demote televisual place to a misrepresentation of the original.

The question which opened each survey, "Please list 5 words you would use to describe Orange County," however, did not make a distinction between televisual Orange County and "reality" because it asks for descriptions of Orange County prior to any television reference in the survey. Responses from Midwesterners and fans clearly describe televisual Orange County, using words such as warm, sunny, wealthy, materialistic, and glamorous. Orange Countians responded to this question with words reflecting everyday lived experiences: similarly warm and sunny, but also boring, suburban, clean, safe, diverse, and crowded. Of course, all of these responses are subjective in nature as this question asked for a simple description of Orange County. However, the Orange County control group describes aspects of the locale, while Midwesterners and fans describe Orange County very differently, with words that better reflect the televisual place. The differences between responses from Midwesterners/fans and residents demonstrate evidence of a changed worldview. For these respondents, televisual Orange County is created at the level of stereotypes which are often put to work by the media as "a process of categorization through which distinctive features … are used to give identity" (BURGESS and GOLD 1985, 9). Stereotypes create place through the association and emplacement of generalized meanings and contribute to a thin, not wholly realized, form of place-making. These types of responses do not fit neatly within comparative or incorporative views, but reside somewhere in between.

Other survey questions, which did not expressly invite comparison, yielded more varied responses, as does the question: "What do you like most about the shows?" Approximately 20% of respondents indicated that they value these shows primarily for entertainment, as a "*mindless*" escape. Examples are: "*Because it is silly and fun to watch*" and "*I like to look at pretty girls and turn my mind off for a half hour*." Such statements do not exhibit a personal attachment to the televisual world or even a belief in it, and can thus be classified as comparative and voyeuristic in nature.

Statements similar to "*I like to see how other people my age are living*" and "*It was fascinating because it was so unrelatable. As an actual housewife, I was mystified by the lives of these women*" describe a relationship with the televisual world that is subjective, yet not fully incorporated into one's own life. Approximately 30% of respondents express such a fascination with seeing a lifestyle that is seemingly so different from their own. After all, the television shows in this study depict extravagant lifestyles, either in affluent coastal communities, or behind the gates of an upscale enclave. These shows purposefully create Orange County as an exclusive entity, a place intended to foster feelings of outsiderness amongst its viewers. The shows are edited and written to focus on the more unflattering aspects of its characters (namely greed, selfishness, and cattiness, which make for good TV), encouraging viewers to "love to hate" this Orange County stereotype and thus to distance themselves from it, and to derive pleasure from the contrast. Contrast works subjectively to reaffirm viewers' own distinct "place" in the world by showing them not what constitutes their place, but some other less-valued place. These respondents engage televisual place to validate their place in the world rather than incorporating televisual place into their lived world. This limited type of engagement works at a slightly deeper level than straight comparison, and does affectively impact subjectivity to create televisual place at a rudimentary, stereotypical level.

The following responses (10% of total) to the same question ("What do you like most about the shows?") can be classified as incorporative in nature and reflect active participation and personal engagement with the places created by television:

It makes me feel like I am just like the cast.

It's fun to imagine that you're a part of it.

I envision myself as a less cool version of Seth Cohen [a lead character].

Rather than outsiders, these respondents directly engage in televisual place-making. As RELPH (1976, 49) describes, "The essence of place lies not so much in [landscapes, cities, and homes] as in the experience of an 'inside' that is distinct from an 'outside' … To be inside a place is to belong to it and to identify with it." Statements like these – being a "part of it" or feeling "just like the cast" – are insider mentalities. Responses such as these show how emotion affectively elides the normative re-presentational divide. Respondents incorporate television's geographic and emotional realism into their own subjectivities, allowing them to engage in televisual place-making at a haptical, corporeal level. It is at this moment that place becomes practice, that place is made relevant and "comes to matter" in our lives (ROSE 2002, 457).

Finally, roughly 40% of respondents cite "drama" as the main appeal of the shows. Although they didn't elaborate or can't explain (e.g. "*Can't stop watching. Drama is addicting. Don't know why. Will never admit it.*"), the lure of melodrama lies in emotional realism. Overt meanings contained in the drama itself do not provide the meaning or value that comes to be associated with a place. Rather, respondents emotionally identify and connect with aspects of the social geographies found in these televisual places. Emotional realism calls forth an affect which evokes a corporeal bond, an experience that nurtures the psyche and gives it a place to rest. Place, here, becomes a "pause in movement" (TUAN 1977, 138), a moment of reflection, a sense that *my* world can be understood through the emotive connections these shows evoke. These emotive connections create meaningful experiences and allow televisual place-making to occur.

A related question focusing specifically on place, "How does the setting contribute to the appeal of the show?" drew a similar distribution of responses. Approximately 45% of responses focus on physical aspects of Orange County which respondents believe are verifiable "truths" of its re-presentation. Statements from Midwesterners and fans such as "*It was always sunny and seemed like a gorgeous place to live*" and "*Because Orange County has beautiful, big houses, and it has a beach, and also the weather*" indicate that more objective respondents enjoy the setting because of elements they know to physically exist in Orange County. Responses from Orange County residents to this question overwhelmingly cited the appeal of beaches and the area's natural beauty. After all, these students live in Orange County and are ensconced in the unmediated version. Understandably, they do not readily identify with mediated Orange County and are experts in comparison.

Additionally, "*the beach*" or "*nice weather*" are not strictly comparative-type responses, but cite selective attributes of *televisual* place. By presenting these physical characteristics to communicate symbolic, generalized meanings, the shows allow stereotypes to overtake Orange County's differentiated and diverse landscape. Here, cultural meaning is naturalized and emplaced in the respondent's mind. Geographic realism works to break down the real/reel divide through the process of naturalization, which creates an "'invisible join' between reality and cultural representations" (DUNCAN and LEY 1993, 4). Similarly, WYLIE (2006, as cited in SHAW and WARF 2009, 1340) notes that "the visual world always transcends our perceptions of it. Our senses can never entirely grasp the richness, complexity, and depth of life." Again, responses which reflect the creation of a televisual place comprised of stereotypes indicate a thin form of place-making. Television has affected subjectivity and modified perceptions of Orange County, but televisual place is neither completely present nor completely absent in the lives of these respondents. Their responses are subjective, falling somewhere between comparative and incorporative. While respondents seek to maintain the "either/or" (either real or unreal) dichotomy, televisual place becomes "both/and."

In answer to the same question ("How does the setting contribute to the appeal of the show?"), some responses fall into the subjective category. For example, Midwesterners and fans answered "*[I'm] fascinated by how "those" people live*" and "*It's one of the trendiest, richest areas in the U.S., so it's a bit like ogling American royalty.*" Unlike those described above, which naturalize cultural meaning, these responses (roughly 45% of total) do not mention physical attributes, but indicate an appreciation for the Orange County lifestyle, as seen from the outside. But as do those discussed above, these respondents also rely on stereotypes as a means to construct televisual Orange County and use what they have seen to fashion a generalized image of Orange County that does not fully connect with their own lives. These respondents engage in thin televisual place-making through a connection with the social, rather than physical, geography of the OC.

The following responses to the above-stated question (10%) illustrate incorporation of the OC within the respondent's life (as ideally envisioned by the melodramatic imagination):

> *The summer setting makes you want to be there.*
>
> *They make you feel you want to live in OC.*
>
> *We all wished we lived there because it is so pretty. We wish we were in the show.*
>
> *[Laguna Beach] is amazing! I would do anything to be there!*

Such responses demonstrate that the meanings communicated by these television shows, through specific production techniques, engender emotional realism and compel respondents to put themselves in this place, to be inside of it (cf. CRESSWELL 2004). As SHAW and WARF (2009, 1338, cf. MASSUMI 2002) note, "affect and representation are indelibly linked." To these respondents, televisual Orange County is directly experienced and felt through the emotional and sensory connections it evokes: it becomes a place as it becomes (personally) meaningful.

The subsequent question "Is Orange County a place you find appealing?" generated a similar mixture of comparative and subjective-type responses from fans. Upon contemplating the attraction of Orange County, participants often cited physically pleasing, stereotypical traits and then projected these characteristics of the television world onto Orange County, concluding that they probably wouldn't fit in:

> *Not after watching Laguna Beach. The people on the show were spoiled brats that didn't appreciate what they had. The County itself looked beautiful though.*
>
> *I do find that it looks beautiful ... but I don't think I like the type of people who tend to live there.*
>
> *Not really, considering you would have to have money to enjoy it.*
>
> *Yes, if I was fabulously rich.*
>
> *I'm sure there's more to the OC, but ... I'd need more density, diversity, and culture.*

These types of responses reveal an Orange County formed by the naturalization of stereotypes: the projection of a generalized identity onto a location results in a less substantial type of place-making, one based upon a non-incorporative engagement with place rather than affective personal connection and participation within it. Finally, although many answered simply "yes" to finding this place appealing, no one overtly addressed televisual OC in an incorporative manner. Rather, opinions were restricted to the locale Orange County is believed to be.

EFFECTS AND AFFECT

Comparative, subjective, and incorporative survey responses coincide with evolving theories in film geography. In the case of Orange County, when televisual place is compared to the locale, it is considered mostly inaccurate by Midwesterners (69%) and OC residents (55%). Televisual place is not to be trusted: "*I don't think it is really like that there*" or "*[It] is probably just for TV.*"

Following the crisis of representation, film has increasingly been seen not as a re-presentation of reality, but as a simulacrum which creates its own reality (CLARKE 1997). Film owes nothing to any underlying truth: "once assembled, a film does not re-represent a world that preexisted it" (DOEL and CLARKE 2007, 897). Film and TV thus construct places where experience is not dependent upon a relationship with the locale that it "plays." The simple fact that *The OC* was filmed mostly in Los Angeles exemplifies this point. Thus, the normative belief that underlies the re-presentational model of film geography often contains a double fallacy. The first is that filming locations do not necessarily correspond to the places they purport to depict (i.e. "crimes against geography" [LUKINBEAL and ZIMMERMANN 2006, 8]), and the second is that re-presentation is a misnomer because TV and film do not re-present that which already exists, but create simulations (MATLESS 1992; DOEL and CLARKE 2007). As MATLESS (1992, 45) describes, "any representation stands as a creative combination and remaking of the world, its status neither below nor above that which has been drawn on." BAUDRILLARD (1994, 6) goes further to posit that simulation subsumes representation:

> Representation stems from the principle of the equivalence of the sign and of the real … Simulation, on the contrary, stems from the utopia of the principle of equivalence, *from the radical negation of the sign as value*, from the sign as the reversion and death sentence of every reference. Whereas representation attempts to absorb simulation by interpreting it as a false representation, simulation envelops the whole edifice of representation itself as a simulacrum.

By expressing distance or self-distinction from the televisual world, subjective survey responses accept the normative belief of the re-presentational model. However, through the application of stereotypes, they exhibit partial acceptance of television's ability to naturalize cultural meaning, a thinly-felt affect that denies the full absorption of re-presentation by the simulacrum. The subjective effects of television, for example, have been cause of much speculation in the pop culture sphere. In chapter three, we presented a great deal of evidence which supports television's tremendous impact on the identity of Orange County, such as the quote from this columnist:

> *The OC* has been a boon to its hometown, culminating in the County's transformation from Los Angeles' ho-hum neighbor to a trend-makers perched on the endless Pacific (VILLARREAL and POWERS 2006).

Subjective-type survey results display a similar effect on respondents' perceptions of Orange County. Seventy-two percent of Midwestern and 85% of fan survey participants stated that at least some of their knowledge of Orange County came from television. Several of the responses presented here illustrate that these television shows have had a significant impact upon the way they think about the "real" Orange County. Answers to a question which ask participants to elaborate on how much of their knowledge about Orange County derives from the shows are exemplary of a subjective take on the televisual world:

> *When I think about OC I think rich white women.*
>
> *I know Orange County like I see it in The OC, with the beautiful houses and the beach, nice cars, hot surfers.*
>
> *I would say that the show has shaped my view, particularly regarding the priorities and characterization of people living in that area.*

Incorporative responses, however, reflect not only a "belief that we can see the real in cinematic representation" (AITKEN and ZONN 1994, 20), but exhibit identification with and participation in the televisual world. Incorporative responses elucidate the value of conceptualizing film as not just a "sign of reality" (AITKEN and ZONN 1994, 20), but as producing a "reality-effect" (DOEL and Clark 2007, 897). For some respondents, televisual place resonates with a sense of belonging, an attachment forged through emotional and personal experience that allows them to "make place" and inhabit it as their own.

Although it is important to note that not all viewers experience televisual place in an incorporative manner, those presented above indicate that televisual Orange County, like other places, is real and directly experienced, as a "meeting place" (MASSEY 1994, 154). FISKE (1985, as cited in MORLEY 1989, 23) describes the meeting place provided by television:

> … the reader … make[s], out of the program, the text that connects the discourses of the program with the discourses through which he/she lives his/her social experience, and thus for program, society and reading subject to *come together in an active, creative living of culture in the moment of reading* (emphasis added).

Television programs use several techniques to induce an affective response which amplifies the emotional connection between people and place. Affect, as defined by CARTER and MCCORMACK (2006, 234), "is a kind of register of sensible intensity emerging from the relations between a multiplicity of processes, corporeal and incorporeal." Mediated places are made through the intersection of a television show and a viewer's life, and these connections build upon each other through weekly engagements, to become places that are continually formed and shaped. SHOUSE (2005) explains that these sensations can

> 'mean' more to people than meaning itself … the pleasure that individuals derive from music [the clearest example of affect] has less to do with the communication of meaning, and far more to do with the way that a particular piece of music moves them … Every form of communication where facial expressions, respiration, tone of voice, and posture are perceptible can transmit affect …

Indeed, the filmmaker EISENSTEIN (1987, 218) opines that "landscape lies the closest to music" when it comes to the "embodiment of the sphere of pure emotionality." The combination of the optical and haptical affects of emotional and geographic realism in *The OC, Laguna Beach*, and *The Real Housewives* congeal in a corporeal experience that engenders televisual place-making. Just as ROSE (2002, 465) urges readers to "move away from unveiling the landscape's meaning in order to better grasp its meaningfulness," SHOUSE (2005) concludes that "the power of many forms of media lies not so much in their ideological effects, but in their ability to create affective resonances independent of content or meaning." In other words, it is not through outwardly apparent meaning contained in the storyline that televisual place-making occurs, but through an affect which produces a sensory, emotive bond to televisual place.

Affect in film, as expressed by LUKINBEAL and ZIMMERMANN (2008, 21) "moves beyond reflect(ion) to engage the content and form of the visual, as well as our own subjectivity, beyond the reality-effect that film affords." KENNEDY (2008, 189) similarly finds that "the emotions aroused in our experiences in viewing movies have a significant role in the creation of our life-worlds ..." KENNEDY (2008, 203) goes so far as to assert that "through film I was introduced to many worlds and realities that ... appear to have influenced my life at an unconscious level far more than I had dreamed." Television programs are designed to induce emotional realism that forges a connection between viewers' lives and the television world. This connection evokes an affective response within viewers which culminates in the emplacement of felt value and ethereal meaning. In these ways, viewers are present in televisual place and engage in televisual place-making. As several survey responses demonstrate, televisual Orange County is not just a made-up place to be forgotten once the show is over and the TV is turned off, but fully exists in the (un)consciousness of viewers and, in this way, is participated in directly. TV's emotional realism "takes us on a voyage through ourselves ..." (LUKINBEAL and ZIMMERMANN 2008, 21) and provides our "affective transport" to place (BRUNO 2007, 40). Our journeys to televisual place are meaningful and directly experienced as they shape ourselves and the world as we know it.

The next chapter addresses the issue of touristic participation in televisual place. Film-induced tourism is the ultimate expression of personal meaning and emotional intensity as experienced through the media, and such attempts to visit a televisual place in person represent a desire to make this experience a more tangible one. Situating televisual place within a locale, however, is an impossible task which often ends in disappointment. Upon arrival, TV tourists are forced to confront the paradox of two places sharing one location. This paradox is addressed in the next chapter.

CHAPTER 7

WHEN WORLDS COLLIDE: THE PARADOX OF TV TOURISM

The Orange County presented to televiewing audiences is exquisite in its beauty and quite seductive in its composition of sunshine, terraced mansions, and trips to the spa. It is also very dramatic and its social spaces are super-charged with affective resonance. Upon visiting Orange County in person, however, no matter how beautiful one might find its physical environment, how spectacular its mansions, or how much money its residents might throw around, direct experience in a locale cannot be equated with mediated place experience. The Orange County we see on TV is, after all, not true life, but true simulation. Although the nature of experience in both is essentially the same (both are real and direct), they produce dissimilar affects and effects. When a fan visits Orange County in search of "the OC," the resulting collision is that of two places in one location.

HYPERREALITY

Film and television images do not reflect an underlying truth or represent a reproducible reality; instead, they create their own (AITKEN and ZONN 1994; CLARKE 1997; DOEL and CLARKE 2007). Moreover, television's reality is hyperreality, a more-than-real or "realer-than-real" world, modeled not on an original, but in the image of image (BEST and KELLNER 1991, 119; cf. BAUDRILLARD 1983). It is a real "retouched and refurbished," often offering a more stimulating experience than the alternative (BEST and KELLNER 1991, 119; BEST and KELLNER 1997). Such hyperreal circumstances are best illustrated by ECO (1983) when his *Travels in Hyperreality* take him to (none other than) Orange County. ECO (1983, 44) observes of Disneyland's Jungle Cruise ride:

> When there is a fake ... it is not so much because it wouldn't be possible to have a real equivalent but because the public is meant to admire the perfection of the fake ... In this sense Disneyland not only produces illusion, but – in confessing it – stimulates the desire for it; A real crocodile can be found in the zoo, and as a rule it is dozing or hiding, but Disneyland tells us that faked nature corresponds much more to our daydream demands ... Disneyland tells us that technology can give us more reality than nature can.

And so, as does Disneyland, TV presents more thrilling, more dramatic, more spectacular, more enthralling places than we encounter in everyday life. SADLER and HASKINS (2005, 197–198) argue, for example, that several Giuliani-era television programs set in Manhattan, including *Seinfeld, Friends*, and *Sex and the City*, offer viewers a postcard-like, tourist-friendly version of the city, a "clean, safe hub of entertainment and consumption" such that with time, "the recurring metonymic representation of the city makes people forget about the real New York" (Figure 67). Similarly, Orange County is commodified and amplified as a succession of beachscapes, cocktail parties, and conspicuous consumption (Figure 68). *The OC, Laguna Beach*, and *The Real Housewives* present a hyperreal Orange County which does not aim to reproduce, or even to misrepresent, the locale of Orange County. Rather, it creates a simulated Orange County, or "the OC." Television has bestowed this hipper nomenclature upon its simulation which, significantly, has never been part of its residents' vernacular. "The OC" signifies a hyperreal place with its own landscapes, icons, values, residents, and associated emotions. Immersion in the simulation – crying alongside the characters as romantic relationships are ended (to musical accompaniment), basking in the same sun at a gorgeous beach – provides an affective, emotional, expertly framed experience in Orange County, much more so than do the mundane activities of driving down the 405 freeway or getting groceries in Newport Beach.

Figure 67. Cast of Sex and the City *(1998)*

Figure 68. Cocktail party in The OC

I'LL TAKE "THE OC"!

Offering more scenery, more spectacle, and more drama than any location, the hyperreal OC looms large in the melodramatic imagination. Survey respondents take palpable pleasure in describing a place they either love to hate or love to love. Some of the more colorful descriptions from the Midwestern and fan survey groups include:

> *Plastic (both in body and credit cards), beautiful facades (of both house and body), but inside things may not be that great.*
>
> *They project people who are very rich, women who spend their husbands' money, and kids who spend their parents' money, spend lots of time at the beach.*
>
> *An oceanfront paradise where everyone has sold their soul.*
>
> *The OC was portrayed as a modern day Eden or a utopian village of wealth and lack of responsibility.*
>
> *A glamorous sun-drenched paradise. Everybody has everything they want.*

It is no wonder, then, that fans often feel compelled to visit a location they've seen on TV or in the movies, "to put themselves in the physical place" of the action (MACIONIS 2004, 91). Film-induced tourism is a well-documented phenomenon (RILEY and VAN DOREN 1992; TOOKE and BAKER 1996; RILEY et al. 1998), and a wide range of push/pull factors bring tourists to filming locations (MACIONIS 1994). In its most extreme form, emotional involvement with a mediated place is so strong that "media pilgrims" embark upon "value-affirming journeys" to fulfill self-actualization motivations (COULDRY 1998, 102; MACIONIS 1994). As BRUNO (2007, 39) observes, filmic place is "charged with layers of topophilic *emotions*, it is invested with the ability to nourish the self. This psychic process involves making claims and demands on the site." For example, COULDRY (1998, 97) finds that visitors to the set of *Coronation Street*, Britain's longest-running soap, are thrilled to find themselves, at last, "on the 'actual street' … where they 'actually film.'" COULDRY's (1998) interviews illustrate (Figure 69):

> *I couldn't believe that I was actually on the spot of the 1983 showdown that rocked the nation, between Ken, Mike, and Deirdre ... I then carried on walking and approached the spot where Ena and Elsie bickered for all those years and even nearly came to blows* (98).
>
> *From the moment I put my foot on the Street I feel like a star. I start my walk down the Street ... I look through ALL the windows and through ALL letter boxes. I touch the stone cladding of number 9. I feel so happy and trouble free when I walk down the Street ... It is the best thing and most wonderful thing I have ever done* (103).

Figure 69. Coronation Street *(1960)*

These motivations further validate ANG's (1985) concept of emotional realism as integral to viewers' connection with televisual place. Film-induced tourism, as an actual enactment of emotional identification with the world of TV, in anticipation of that magical moment when the normally out-of-reach mediated world intersects our ordinary one (COULDRY 1998), is the ultimate expression of the irresistible attraction of mediated place. Film-induced tourism can be understood as an attempt to procure the haptical, or anti-optical, experiential meaning that emanates from mediated place.

In our hyper-mediated world, there remains a quixotic drive to see the "original," as evidenced by the popularity of art museums or visits to a set or soundstage, to see for ourselves the actual artwork or actual site where filming occurred. In the case of hyperreal televisual place, TV tourists soon realize that the "original" does not exist, but is a myth (LUKINBEAL and ZIMMERMANN 2006). Television excels in composing and simulating a sense of place, but place, as CRESSWELL (2004, 11) explains, is "not just a thing in the world but a way of understanding the world."

For iconic, symbolic locations with a vivid representational legacy, difficult can be the task of living up to the spectacular standard of their "cultural mythology" (STENGER 2001, 61). Not far from Orange County is Hollywood, the place perhaps most idealized in the American psyche and worldwide as center and symbol of the motion picture industry. Hollywood Boulevard, the physical location, is far from our idea of Hollywood from the movies. Since its decline into seediness, the Hollywood and Highland area is now being resurrected as a simulacrum of Hollywood's Golden Age, the focal point of development being a shopping complex designed to look like the iconic movie set, Babylon Court (STENGER 2001). This all to create an experience more satisfying for tourists, a space where "the 'Hollywood' of the imagination may be tangibly validated and consumed" (CURTI et al. 2007, 63). But, as CURTI et al. (2007) note, Babylon Court and its selection of chain stores offer shopping opportunities like any other. The question remains, does even a newly redeveloped "Hollywood" lose its luster when you purchase a pair of jeans at Hollywood and Highland that you could have more easily found at the mall back home (Figure 70)?

Figure 70. Babylon Court at Hollywood and Highland

JACKSON (2005, 186, citing CROUCH and LÜBBREN 2003), an Englishwoman, relays her disappointing visit to Los Angeles. In her words:

> The stars' homes were fascinating, solely because we found them architecturally disappointing … Visiting the Chinese Theatre brought the realization that whenever this location is filmed … we see the stars, the red carpet, and the adoring crowd. Yet, immediately adjacent was wasteland, fronted by old railings sporting tacky advertisements, a bleak contrast to the TV-transmitted glitzy glamour … We made sense of most places by comparing them to screen images, being surprised at our own naivety, having failed to appreciate how extensively our imaginations had been fueled by the "metaphoric construction of ideas and desires."

A *Vanity Fair* columnist ridicules the *Sex and the City* bus tour for a similar reason:

> These women on the bus are missing the point. The storyville they're looking for doesn't exist and never did, and trying to search for the literal in literature inevitably kills the object of affection, murders the fiction stone-dead … After four hours we're dropped off in Midtown. The bus is strangely subdued. The women file out swiftly, each tipping our stand-up guide a grudged, crumpled ten without making eye contact. We all realize what an obsessively ridiculous, embarrassing, empty, and needy exercise this has been. I can't imagine anyone here would repeat the trip, or recommend it to a friend. And I guess that a good many of us will keep it as a guilty secret … As if to underline our gullibility, they deposit us beside the HBO store (GILL 2009, 53).

Another visitor to the set of *Coronation Street* reflects upon the experience:

> It's like when you were a child, you imagine something, then you go back to it as an adult and it's totally different … it was exactly like that. Everything just seemed very small and flat (COULDRY 2005, 64).

While "the OC" is anything but boring, Orange County survey respondents confirm that a life lived there shares everyday characteristics with everywhere else:

> *The always exciting and fun aspect isn't true; we're just like everyone else.*
>
> *My life isn't as dramatic nor as exciting ...*
>
> *It doesn't show how regular kids live in a house that's three bedrooms or apartments.*
>
> *We have lives and jobs, real jobs where you actually have to work, and we drive normal cars.*
>
> *For some reason when it's on TV it seems to be more glamorous.*

TORCHIN (2002) uses the Manhattan TV Tour to describe how film can imbue everyday sites with personal significance; however, as these experiences illustrate, when comparing mediated place to the spot at which a tour bus has dropped you, this supposed "original" appears far more ordinary. When expected to provide the glossy images of true hyperreality, even in Orange County, absolute location intrinsically fails to measure up. JACKSON (2005, 194) well-expresses this sentiment, "Our visit to LA demonstrated that when the two collide, disappointment may ensue." By visiting a location with hopes of experiencing the beauty, glamour, and "aura" of mediated place, a curiosity may be satisfied, but essentially, we seek a place that is impossible to find. When we arrive at a site expecting to be enveloped and stimulated by the place we helped to create, we find it absent of affect. Emotional, imaginative, resonate televisual place collides with a location lacking this intimate meaningfulness, and we "inevitably kill the object of *affect*ion" (GILL 2009, 53; emphasis added).

The idea that cinema and television re-present ontologically stable objects is killed because of the "radical negation of the sign as value" (BAUDRILLARD 1996, 6) – the negation that an absolute location is an equivalent referent to its representation. The re-presented objects, events, and narratives of television and cinema are rarely, if ever, a "sign of reality" (AITKEN and ZONN 1994, 20). The sign is sentenced to death because the simulacrum of television and cinema frees the signified location from its absolute referent. Representation kills affection by limiting the sign to reality, whereas the absent/presences of a simulated place at a filmed location requires the referent to be left open and relative to the affective bond individuals have created with the televisual or cinematic place.

Television's landscapes are more scenic than a drive on the 405 freeway and its emotional realism evokes more feelings than a day spent shopping at one of its malls. Television brings together visual, emotive, musical, and other elements to create places in which meaning is more concentrated, experience is more engaging, and the place-making process more imaginative. Televisual place is, in a word, *more*. To paraphrase ECO (1983), television tells us that its places better correspond to our daydream demands. When someone who has affectively engaged in televisual place-making visits Orange County in search of the mediated place of the same name, he or she is unexpectedly forced to reckon with the paradox of TV tourism. At the very location where we had hoped to re-experience our televisual place in tangible form, we instead encounter lived regularity. We might find ourselves eating lunch at the very spot where two *Laguna Beach* cast members ate lunch on the show, hoping for a fulfilling experience at the "actual" location, but the screen's spectacular frame, edited montage, emotive music, and exciting drama are gone. In their place might be noisy traffic, a bad view, and a long wait for food. Lunch here is just lunch, but when experienced alongside the characters, within the hyperreal televisual world, we are emplaced within its social geography, and lunch becomes much more.

THE OC EXPERIENCE

Television exposure has indeed been a boon to tourism in the region, and many fans make the pilgrimage to participate in the "OC" lifestyle they have experienced on TV. Although neither the Newport Beach Convention and Visitors Bureau (NBCVB) nor the Laguna Beach Visitors and Conference Bureau (LBVCB) have formal mechanisms in place to track TV-induced tourism, officials from both organizations feel that visitorship has definitely increased as a result of these shows and can cite significant increases in hits to their websites (BIJLANI 2008; LEONCIO 2008). Area newspapers provide anecdotal evidence of a major tourist influx:

> Shopkeepers are swamped with tourists and callers wanting to track down the stars. 'I've spent the last two years saying "Stephen [*Laguna Beach* heartthrob] doesn't work here,"' said Thalia Street Surf Shop owner Pam Cocores (SMITH 2005).

> People have gone to the high school looking for *Laguna Beach* characters. Out-of-town teens have tried to enroll in the hope of being discovered. There have been minor traffic accidents as motorists gawked at the filming around the city (KELLY 2006).

With the exception of establishing shots, *The OC* was not filmed in the County of Orange. Acutely aware of this problem, the NBCVB created an "As Seen on *The OC*" map to appease star-struck tourists, directing visitors to the spots which inspired many of the locations used in the show (Figure 71). Guests are also encouraged to visit a site (ironically, the front office of an RV park) where handprints of *The OC* cast members can be viewed (Figure 72). Explains a NBCVB official:

> … even though they didn't film here, showing them around the destinations gives them a taste of the lifestyle, because that's what the show is selling, really, is the lifestyle, the destination, and you still get a feel for that, no matter if you're really in front of the Cohen house (LEONCIO 2008).

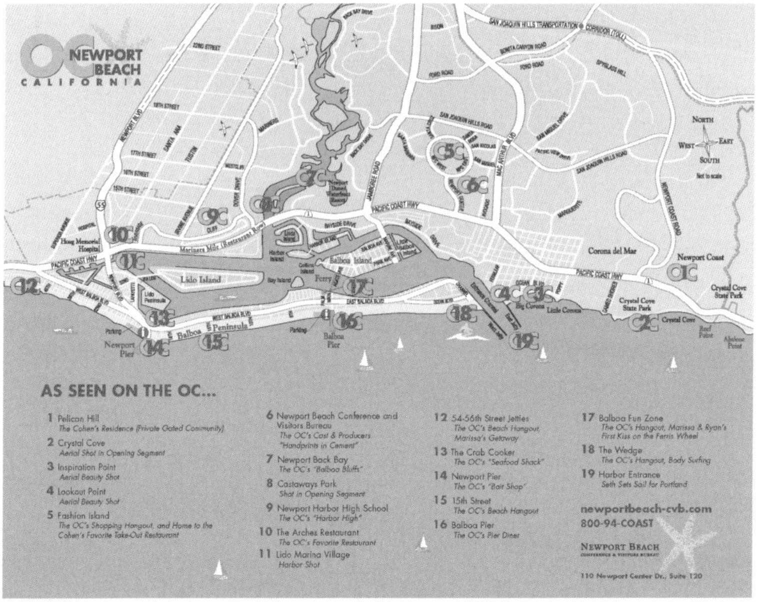

Figure 71. NBCVB's "As Seen on The OC" *map. Available from www.visitnewportbeach.com*

Figure 72. Handprints of The OC *cast as displayed at Newport Dunes Waterfront Resort*

Laguna Beach is, of course, filmed on location, but visitors who want to see actual homes of the cast members will be disappointed, as most reside in gated communities that are well beyond the reach of the general public. Laguna Beach High School, which we presume cast members attend (the show contains little mention of school), is easy to find, however, and has become a busy intersection of star-seekers (KELLY 2006) (Figure 73). The LBVCB, for its part, offers visitors a list of filming locations to assist in their search for "the real Orange County," including shops and eateries that cast members frequent on the series. In the opinion of Judy BIJLANI, Executive Director of the LBVCB, "there might be a level of disappointment not seeing stars walking down the street, but this is mitigated by being here, being submersed in the experience" (BIJLANI 2008). The general industry sense is that TV tourists leave Orange County feeling satisfied, because it has so many other attractions to offer.

Figure 73. The (in)famous Laguna Beach High School

While *The OC* craze has died down in the U.S., the show is in syndication abroad and Newport Beach now welcomes many Japanese and other international fans of the show (LEONCIO 2008). To take advantage of the area's newfound popularity with foreign visitors (most of whom also go to Disneyland), Coach USA/Gray Line, since 2005, has offered "The OC Experience" bus tour. Says Belinda TRANI, Director of Sales for the tour, approximately 90% of tour-takers hail from abroad, mostly Canada, Australia, and New Zealand, and most are, in fact, fans of *The OC,* "looking to see where these people may shop, eat, and play. The international visitors are very fascinated with the California lifestyle" (TRANI 2008).

On Tuesday, December 23, 2008, one of the authors (Ann) took the OC Experience tour. Three days per week, the tour departs from Anaheim (Disney central) and visits a handful of Newport Beach points of interest. Along on the tour were a young family from Indiana and two couples in their twenties or early thirties. One couple was from Canada, the other, Australia. Each group had come to Orange County specifically to visit Disneyland and other area theme parks, but were lured to this tour by promise of its proximity to the beach and opulent mansions, as advertised in the Southern California Gray Line promotional brochure:

> Escape for a day to world renown Newport Beach, crown jewel of the California Riviera and famous for its fabulous harbor, incredible beach front mansions and world class shopping. See where John Wayne, Humphery Bogart [sic], Tyrone Power, Buddy Ebsen and many of Hollywood's finest used to live. Today, Newport Beach continues to be the residence and playground to many sports celebrities, film and television stars and serves as the "on location" site for Fox Televisions [sic] hit series *The OC.*

Our tour began on the freeways as the bus made its way to the coast, accompanied by commentary from the tour guide (a long-time Orange Countian, possibly prone to exaggeration). Our first stop was a brief photo opportunity at a small park overlooking Newport Harbor, followed quickly by another short photo stop at the beach near Newport Pier. We proceeded to Balboa Pier and were dropped at kitschy, but not quite open yet, Balboa Fun Zone to spend an hour awaiting the start of our Harbor Cruise (Figure 74). The boat's captain was rather late to arrive, but as promised, we saw many oceanfront mansions and learned of notable persons who resided in them (Figure 75).

Figure 74. Balboa Fun Zone, as seen from the Harbor Cruise

Figure 75. View of Newport Harbor from the dock

After the cruise, we met back at the bus at the appointed time and our tour continued with a drive through the exclusive Newport Coast McMansion development (Figure 76). Although no one on the tour this day was a fan of *The OC* (except me) and thus no one on the tour could be considered a television-induced tourist, our tour guide did point out a couple of (very possibly fraudulent) filming locations from *The OC*. He even pulled the bus over so that we could get out to take a picture of a home supposedly used in the series.

Figure 76. Newport Coast McMansions

Our final stop on the tour was Fashion Island Mall, a fancy indoor/outdoor shopping complex, at which we were dropped for at least two and a half hours.

The tour had arranged for our group to be greeted by the mall concierge, who presented us with a gift bag containing a nice souvenir coffee mug, perhaps to reassure us that our trip to the mall was somehow special. Upon getting back on the bus, one tour-taker commented to his wife and daughter that Fashion Island was just about like their mall back home, but with some more expensive stores (Figure 77).

Figure 77. Newport's Fashion Island Mall

I asked our tour guide about the frequency, in this post-OC period, of TV-induced tourists on the OC Experience tour and he replied that about half of his guests are fans of the show. He tells them "This is not *The OC*, but a tour of Orange County; otherwise you see a hospital, roads … that's why I insert so much of myself into the tour …" Underwhelmed by the Orange County experience I had just purchased, I understood the necessity of this disclaimer. For that day, Newport Beach lacked the magic that I and other of the media-influenced hoped the sights would evoke. Instead, it seemed a bit pathetic for our small, rather unenthusiastic group to be shuttled around on a large bus to take a few photos and go to the mall. Although the cruise was scenic and the mansion views impressive, overall, the tour provided a surface-skimming visit to what seemed to be a somewhat regular place. Newport may be the "crown jewel of the California Riviera" (Gray Line brochure) but even so, I never felt too far from the regular world of malls and freeways, motels and tourist traps of the unmediated Orange County. Several survey respondents who had visited Orange County felt the same way:

> *I didn't see too much of the glamour. It was pretty chill and I didn't see too many fancy houses or anything of the sort.*
>
> *Not as many rich, gorgeous people walking around ... Lots of nice cars though in restaurant parking lots.*

Much less lavish and more normal than what is shown on TV.

Way more traffic and cloudy gray weather.

Besides being more smoggy, I guess the people were more ugly and fat than in the shows.

Orange County cleaned up nicely for its prom photo. It's a beautiful place in real life, but not nearly as "perfect."

It is difficult, indeed, for a mere location to live up to the image presented by its mediated counterpart, especially when that place is televisual OC, an impossibly perfect beachfront setting which produces a powerful affect. Perhaps we are better off staying home, where televisual Orange County is safe from being "thinned out" by exposure to the hum-drum experience at a physical location. Televisual place derives its hyper-meaningfulness, or "excess of meaning" (BAUDRILLARD 1994, 28), and its charm from the stirrings of affect. Without TV's affect, we only get the location, which must become meaningful in different ways, through different processes.

This chapter has positioned "the OC" as a hyperreal simulation, providing a very stimulating, engaging, and affective experience of place which is only available through television. Touristic experience suggests that searching for televisual place in specific locations will likely end in disappointment. Regardless, beyond personal affect, televisual place produces powerful effects which resonate within greater Orange County.

CHAPTER 8

CONCLUSION: "RIGHT BACK WHERE WE STARTED FROM"

Well before the TV shows examined in this study aired, Orange County had given America its first theme park (Knott's Berry Farm) and its most famous (Disneyland), its first drive-in church (Robert Schuller's) and one of its biggest mega-churches (Saddleback). It supplies the U.S. Olympic team with athletes in events requiring copious amounts of sun and water: swimming, volleyball, softball, and water polo (REID 2008). It sports the newest of new towns, the most beautiful of beaches, and the most luxurious of shopping malls. Its mode of development is so emulated that there is now an Orange County, China full of "replicas of Southern California homes" built in the image of the American Dream, just north of Beijing (ROSENTHAL 2003) (Figure 78). Fittingly, SOJA (1996, 238) offers this description of the many wonders of Orange County:

> … you can have *anything* you want in iconic Orange County, where every day seems just like yesterday but where the ever-present *Now*-ness of tomorrowland makes the *Then* hard to find; where every place is off-center, breathlessly on the edge, but always right in the middle of things, smack on the frontier, nowhere yet now/here like home. To its avid promoters, Orange County is a theme park-themed paradise where the American Dream is repetitively renewed and made infinitely available: as much like the movies as real-reel life can get. This resplendent bazaar of repackaged time and spaces allows all that is contemporary (including histories and geographies) to be encountered and consumed with an almost Edenic simultaneity.

Figure 78. Orange County, China

In the production and perfection of image, Orange County is unmatched. As SOJA (1996) intones, Orange County is as close to the hyperreal places on our TV and movie screens as an actual place can get. Indeed, *The OC, Laguna Beach*, and *The Real Housewives* would seem the perfect complement:

> Perhaps more than any other place, more than the desert, more than either New York or Los Angeles, Orange County verifies Baudrillard's iconoclastic vision of the anesthetic enchantments of hyperreality. It is the generative utopia, a make-believe paradise that successfully makes you believe in make-believing, the most irresistibly California-looking of all the Californias, the most like the movies and TV "situation comedies," the most like the promised American Dream (SOJA 1996, 274).

Regardless of the magical qualities of its iconic built environment, "The OC" is much less diverse, much wealthier, and decidedly more dramatic than the Orange County that three-million-plus persons call home. No matter how enchanting the area might seem to a Disneyland guest from the Midwest in the winter, it also has traffic, poverty, and bland housing developments – all of the trappings of a regular life lived in postsuburban America. Although televisual Orange County and its locale provide vastly different experiences to visitors, "Orange County" is by no means exclusive of "the OC."

Recall how HANNA (1996) describes the relationship of television's Cicely to Washington's Roslyn. Roslyn's identity is bound up in its re-presentation as Cicely and all other representations that came before it. Just as televisual Cicely is now part and parcel of Roslyn, so "the OC" – the landscapes, emotions, and social spaces that comprise it – is an undeniable component of greater Orange County. Viewers cannot help but project the symbolic meaning and meaningfulness of televisual place onto it. Orange County, is always-already a "combination of the material and the mental and cannot be reduced to either" (CRESSWELL 1996, 13).

LIKE IT OR NOT …

Residents of Laguna Beach and Coto de Caza uniformly complain to reporters about Orange County's television-induced image. Their comments reflect an acute awareness of how TV *is* reality. Laguna Beach, once a quaint art colony known for its thriving gay community, is fast becoming an affordable residence for only the super-rich, a transition *Laguna Beach* epitomizes and even exploits (SMITH 2005; BLY 2006; KELLY 2006). Opines one resident: "This is not depicting our town the way it should be depicted. Our town is about art and culture, and school should be about education" (STEINHAUER 2006). And another:

> We are an affluent, tolerant community with a long heritage of culture, character, and art. The MTV show has hijacked our identity, and people are beginning to feel violated. It has become a metaphor for everything we don't want to be (BLY 2006).

Laguna's former mayor (then current), Cheryl Kinsman, chimes in on a more positive note: "I think [the show] will make us look more desirable. It could do for us what *90210* did for Beverly Hills. It will glamorize Laguna Beach" (RITCHIE and ZLOTNIK 2004). These contradictory remarks speak to the broader socio-economic transition the region has undergone since the 1990s (see chapter three).

Local columnists who enjoy dishing on *The Real Housewives* express a general distaste within the Coto de Caza community for its newfound infamy:

> This popularity is all much to the dismay of the residents of Coto de Caza, most of who just wish we were not associated with the self-indulgent, materialistic, and ridiculous people portrayed on the 'reality' show. I've got some bad news for those Coto folks – we're famous (SICKLER 2007a).

> Since *The Real Housewives* has been on the air, most of the country has learned almost everything they know about [Coto de Caza] from the TV show. Which, for Coto residents, is really scary (SICKLER 2008).

Stephen SICKLER (2007b), a local columnist, describes a dinner conversation with Real Housewife Jeana:

> "Jeana, most Coto residents really hate the show ... They think it makes all of us look so bad to everyone. One guy wants to start a class-action lawsuit for defamation of character." She shook her head in disbelief as her phone rang. Her husband was back in jail for drinking ... More reality to come.

In fact, *The OC*'s negative characterization of Chino ("Chino, ewww!" squeals a snobby Newport Beach teen upon learning Ryan's origin) and derogatory remarks made by its characters about neighboring Riverside (Wife: "He basically called me white trash. He said I was from Riverside!" Husband: "Honey, you *are* from Riverside.") have been perceived as so libelous to have prompted threat of litigation from a Riverside councilman (STRAIGHT 2003; LEDUFF 2005; VILLARREAL and POWERS 2006).

Travel literature also engages in playful attempts to separate Orange County from its memorable televisual counterpart, reinforcing the normative belief in representation. A headline in *National Geographic Traveler* for an article on Laguna Beach uses the opening line from *The OC*'s theme song and then discredits televisual place: "California, Here We Come: The Real 'OC': Never mind the television show. Orange County's much better in real life" (LIMSKY 2008, 147). A *Sunset Magazine* article, after quoting the lyrics to *The OC*'s theme song, informs readers:

> Watch the show, and you'd think that Southern California's Orange County is a place where a new BMW, stylish clothes, and a surfboard ... are make-or-break necessities. Perhaps ... but there's more: secluded beaches, kid-friendly tide pools, even some fine art. And even a grommet (an inexperienced surfer) on a budget can experience what one fairly erudite California historian called "the last American paradise." And that's the real OC (LANSING 2004).

Finally, an ad placed by the Anaheim Orange County Visitor and Convention Bureau (slogan: "It's So California") reads: "Be swept away by our reality ... 42 miles of beautiful beaches, 4 seasons of warm weather, 1 real OC." The obvious effort to return ownership of the word "real" to the locale only further demonstrates how material and mediated place are utterly entangled, a revelation realized and embraced by county boosters.

Thanks to the truism "any publicity is good publicity," the "OC" moniker has recently popped up in several public venues, a direct result of the popularity of *The OC* and the name recognition it and the other shows have lent the county (LOWE 2008). There has been a push to incorporate "Orange County" into the name of the area's John Wayne Airport, and Dana Point Harbor has been redundantly rechristened "O.C. Dana Point Harbor." The Orange County Fair and Exposition Center has been renamed the OC Fair and Event Center. The inaugural OC Marathon was run in 2005 and for the second annual race, a cast member of *The OC* was given the honor of firing the starting gun (OVERLEY 2007; OVERLEY 2008; LOWE 2008). Most recently, the popular acronym has been bestowed upon several government agencies, including catchy changes from "Resources and Development Management" to "O.C. Infrastructure," "Integrated Waste Management" to "O.C. Waste and Recycling," and "Harbors, Beaches and Parks" to "OC Parks." Even *Beach Patrol*, a television show documenting "real" lifeguards, has been renamed *Ocean Force: Huntington Beach, OC* for the reason that "'the OC' is synonymous with drama" (GRIFFIN 2008). Residents continue to maintain, however, that "no one except tourists call Orange County 'the OC'" (LOWE 2008).

Of course, the county's tourism promoters are also putting the "OC brand" to good use. As Audrey LEONCIO (2008), a representative from the Newport Beach Convention and Visitors Bureau (NBCVB), explains:

> [I]t's not really portraying us in the best light, [but] it's still capturing the overall essence of the destination. It's still showing the luxury … and that's really how we market the destination …

Representatives from the Anaheim/Orange County Visitor and Convention Bureau explain that "the show [*The OC*] launched us into a new place in the minds of consumers" (CALI 2008). "It's not just Disney, families, and Mickey, but a celebrity status destination filled with beautiful places and people" (CALI 2008). "We couldn't have paid for the exposure *The OC* has given us" (GALLAUGHER 2008).

The Orange County Tourism Council has recently undertaken a new regional branding effort which explicitly uses the term "the OC" and a logo which bears more than a passing resemblance to that of the TV show (Figures 79 and 80). This was a controversial decision, one not whole-heartedly supported by residents, but the tourism council felt they must take advantage of, and even embrace, the name recognition that the show has already provided (FEARY 2008). Says Gary Sherwin, head of the NBCVB, "What other part of the world has three reality shows that take place in it? I think we are really fortunate and need to try to capitalize on it" (OVERLEY 2008). These steps toward seducing tourists with "a fun-and-sun image made famous by Fox's *The OC* and a smattering of reality shows" (OVERLEY 2008) reinforce the fusion of the material and mediated as this singular image of place is projected worldwide. In the words of a Midwestern survey respondent: "The portrayal on TV comes from so many shows with similar themes, it's hard to see [Orange County] any other way." In the end, Orange County is what it becomes on TV.

Figure 79. The OC *logo*

Figure 80. The Orange County Tourism Council logo

"THE SUN SETS ON THE OC"

It has now been seven years since *The OC* debuted and Orange County was introduced to America and the world. Since then, *The OC* has been cancelled, is in syndication worldwide (fans reside everywhere from Turkey to Norway to South Africa to Japan), and creator Josh SCHWARTZ' most recent television hit, *Gossip Girl*, is set in upper-crust, prep-school Manhattan (Figure 81).

Figure 81. Cast of Gossip Girl *(2007)*

Laguna Beach, after an unpopular third season, headed for *The Hills* (Figure 82). This spin-off reality drama follows *Laguna*'s main character, Lauren, to her young adult life in Los Angeles and became, in its own right, a pop cultural sensation. In one episode of *The Hills* (first aired December 8, 2008), Lauren nostalgically returns to Laguna Beach to help her parents move out of the extravagant home with party-ready patio, pool, and hot tub in which we saw her close her high school years. *The Hills*, in December 2008, spun off a duplicate series called *The City*, set in New York City (Figure 83).

Figure 82. Cast of The Hills *(2006)*

Figure 83. Cast of The City *(2008)*

The Real Housewives of Orange County, in 2011, aired its sixth season, a much-anticipated return to the OC after the *Real Housewives* franchise expanded its coverage to wealthy and equally over-the-top ladies in New York City, Atlanta, New Jersey, Washington D. C., Beverly Hills, and Miami (Figure 84). In post-recession seasons five and six, several housewives face shrinking incomes and are scrambling to downsize their too-extravagant homes.

Figure 84. Cast of The Real Housewives of New York City *(2008)*

In a press release announcing *The OC*'s demise (titled "The Sun Sets on *The OC*"), SCHWARTZ eulogizes: "For a certain audience, at a certain time, *The OC* has meant something" (Fox 2007). Indeed it has.

PREVIOUSLY, IN THIS BOOK …

This book is an exploration of televisual place and the processes of place-making through television. It has sought to establish the importance of mediated place as a meaningful experience of and in the world, as a "meeting place" (MASSEY 1994, 154).

Television employs several techniques which work together to create place – places carefully designed to seduce, engage, and affect viewers. The televisual place-making process begins with landscape. Each show commences, not immediately with narrative action, but by first setting the scene. The establishing shot and the montage of spectacular images which comprise the opening credits imbue a program with a sense of place and geographic realism. By describing place, these techniques are essential in encouraging the suspension of disbelief and audience interest in and subsequent enjoyment of the show. Moreover, these broadly-based landscapes – a beach which abuts a dramatic coastline, a golf course flanked by mansions – carry generic cultural meanings and naturalize stereotypes of the social geography of this place. These stereotypes and techniques guide viewers in their journey and emplace them within this hyperreal landscape.

During most of a show, however, landscape functions as a social space. Here landscape provides the spaces – in homes, restaurants, spas – within which the narrative unfolds. Within landscape, camera close-ups show characters' emotions, characters reveal their thoughts and feelings to each other, and characters move through the frame. It is within landscape as space that viewers are more deeply drawn into televisual place and engage with it.

These techniques work with and contra each other throughout a program to create televisual place. There is a continual tension between description and narration, between place and space. Social space is periodically interjected with symbolic landscapes (transitional shots) and these are repeated in each episode and throughout the series to reinforce and re-inscribe sense of place. On the other hand, although they tempt a voyeuristic desire to linger on the view, the forward movement of the narrative exerts a strong attraction, and these spectacular landscapes give way to narrative space. Week after week, as each show airs, viewers revisit this place and, through the portal of landscape, are further drawn into the world that television creates.

It is within narrative social space that we forge an emotional connection to televisual place. The elaborate storylines of melodrama (of which reality TV is also a part) may be unrealistic at face value, but viewers identify with the emotions they elicit. Pent-up feelings find expression in melodrama, and this emotional effect adds an essential dimension to our daily lives. This effect becomes affect as viewers integrate melodrama's emotional realism into their own emotional realities.

The melodramatic genre employs narrative conventions which operate within social space to assist in communicating emotional realism to audiences. Close-up shots of characters' facial expressions, as well as shots which linger on these faces at the end of a scene, and importantly, the use of music, stir up viewers' emotions and affectively transport them to televisual place. Such intensities and sensations work to interweave a character's emotional journey with a viewer's subjectivity.

Televisual place is thus created through production techniques which foster geographic and emotional realism. During the course of a show, viewers continually move between gazing upon large-scale landscapes and emotionally participating in small-scale social space. The interplay of cultural meaning inscribed upon the TV world and the personal meaningfulness of direct experience within it comprises the televisual place-making process. But ultimately, this process is dependent upon viewer engagement.

Survey responses demonstrate several modes of engagement with televisual place. Often, viewers confront televisual place at a fictional level, which reinforces a strict distinction between the viewer's lived world and the world seen on TV. This viewpoint corresponds to a conceptualization of filmic place as a representation of reality. Many other survey respondents exhibit a subjective belief in, and partial acceptance of, the televisual world, indicating that television has changed their worldview. This belief is built upon generalized meanings and stereotypes which act to naturalize expressions of geographic realism and the social geographies presented on TV. These respondents distance themselves from the televisual world and are engaged in a thin form of place-making. Some respondents, however, possess an incorporative relationship with the televisual world, displaying an affective connection which effectively breaks the boundary of the real/reel dichotomy, allowing them to journey into televisual place. These respondents directly participate in televisual place-making; it is a meaningful experience with emotional value.

All of the techniques assembled in a television series work together to create and re-create a visual, emotional, and affective place of meaning and experience. This process constructs, builds, and reinforces place with each episode watched and enjoyed. Televisual place is thus a place like any other. It is a place that becomes personally meaningful, it is a meeting place which exists at the intersection of the multiple social processes that work to construct it, and it is a place that is experienced directly by its audience.

Televisual place, at the same time, is unlike any other. It is hyperreal in character and can provide a more stimulating, more engaging, and more fascinating experience than can be found in most any locale. The mechanisms which create televisual place produce a powerful affective response which captures both the senses and the imagination. So intense can be identification with these hyperreal televisual places that fans are often compelled to make a pilgrimage, to seek out the mythic, auratic "original," but tourists seeking televisual place are confronted with the paradox of two places sharing one location. Where they seek to nurture their psyche with the affective intensity of the hyperreal, they encounter an unmediated filming location which lacks affect. Televisual place is unreachable within everyday mundanity and cannot be defined by absolute location. Televisual place can only be found through the affective transport its production techniques provide.

From fast-growing defense industry suburb to planned community mecca and antithesis of LA, to recently starring in television as one of the most glamorous/notorious counties in the country, Orange County has always been an amalgam of mediated stereotypes and identities. Even before television elevated Orange County in the American and international pop-consciousness, it was, and always will be, a mediated place. *The OC*, *Laguna Beach*, and *The Real Housewives* have exerted a tremendous transformative impact upon the area's popular image, demonstrating that no place is wholly material.

More importantly, this trio of shows has contributed to the creation of a place called "the OC." The OC demonstrates that place cannot be restricted to a location, or even to an image on the screen; place is as much within ourselves. Place emerges as a process requiring active participation and is created as viewers connect with the televisual world. Televisual place transmits affective intensities and carries emotional registers that are part and parcel of meaningful experiences of, and in, the world. We must remember that mediated place—including the televisual—is not about the representation of one world, but the ongoing creation of many.

REFERENCES

ABCARIAN, R. (2006): The Not-So-Sunny Side of the 'Beach,' *Los Angeles Times*, November 3, 2006. http://infoweb.newsbank.com (accessed January 16, 2008).

ADAMS, P. (1992): Television as Gathering Place. *Annals of the Association of American Geographers* 82 (1), 117–135.

ADAMS, P. (2009): *Geographies of Media and Communication*. Oxford.

AITKEN, S. and L. ZONN (1994): Re-Presenting the Place Pastiche. S.C. AITKEN and L.E. ZONN. (Eds.): *Place, Power, Situation and Spectacle: A Geography of Film*. Lanham, Maryland, 3–25.

AITKEN, S. and D. DIXON (2006): Imagining Geographies of Film. *Erdkunde* 60 (4): 326–336.

ANG, I. (1985): *Watching Dallas*. London.

ANTON, M. and J. MENA (2004): The Hard Life – Santa Ana Style, *Los Angeles Times*, September 5, 2004. http://articles.latimes.com/2004/sep/05/local/me-santaana5 (accessed November 14, 2010).

ARELLANO, G. (2008): *Orange County: A Personal History*. New York.

BANERJEE, M. and P. MARX (2008): Ally Lives Just Next Door ... German-U.S. relations in Popular Culture. LUKINBEAL, C. and S. ZIMMERMANN (Eds.): *The Geography of Cinema – A Cinematic World*. Stuttgart, 155–170.

BARNES, T. and J. DUNCAN (Eds.) (1992): *Writing Worlds: Discourse, Text and Metaphor in the Representation of Landscape*. London.

BAUDRILLARD, J. (1983): *Simulations*. New York.

BAUDRILLARD, J. (1988a): *Jean Baudrillard: Selected Writings* (edited and introduced by Mark POSTER). Stanford, California.

BAUDRILLARD, J. (1988b): *America*. London.

BAUDRILLARD, J. (1994): *Simulacra and Simulation*. Ann Arbor.

BELLAFANTE, G. (2006): The New Mrs. Robinsons, on Bravo and MTV, *New York Times*, June. 4, 2006. http://www.nytimes.com/2006/06/04/arts/television/04bell.html (accessed November 14, 2010).

BELLAFANTE, G. (2007): A Teenager in Love (So-Called), *New York Times*, October 28, 2007. http://www.nytimes.com/2007/10/28/arts/television/28bell.html (accessed November 14, 2010).

BENJAMIN, W. (1968): *Illuminations*. New York.

BEST, S. and D. KELLNER (1991): *Postmodern Theory*. New York.

BEST, S. and D. KELLNER (1997): *The Postmodern Turn*. New York.

BIJLANI, J. (2008): Personal communication, February 21, 2010.

BIRD, C. (2008): All-Vietnamese Scout Group in Irvine Grows to 250, *Orange County Register*, July 28, 2008. http://www.ocregister.com/articles/scouts-huong-vietnamese-2107859-kids-pham (accessed November 14, 2010).

BLUSH, S. (2001): *American Hardcore: A Tribal History*. Los Angeles.

BLY, L. (2006): The Real Laguna Beach Disdains its MTV Image, *USA Today*, March 2, 2006. http://usatoday.com/travel/destinations/2006-03-02-laguna-beach_x.htm (accessed November 14, 2010).

BRADSHAW, M. and E. STRATFORD (2000): Qualitative Research Design and Rigour. HAY, I. (Ed.): *Qualitative Research Methods in Human Geography*. Oxford, 37–49.

BROOKS, D. (2001): Castle in a Box, *The New Yorker*, March 26, 2001. http://www.new yorker.com/archive/2001/03/26/010326fa_fact_brooks (accessed November 14, 2010).

BROWNFIELD, P. (2006): "Housewives" in a Parallel Universe, *Los Angeles Times*, March 21, 2006. http://articles.latimes.com/2006/mar/21/entertainment/et-housewives21 (accessed November 14, 2010).

BRUNO, G. (1992): Streetwalking Around Plato's Cave. *October* 60 (2), 110–129.

BRUNO, G. (2002): *The Atlas of Emotion: Journeys in Art, Architecture, and Film*. New York.

BRUNO, G. (2007): *Public Intimacy: Architecture and the Visual Arts*. Cambridge, Massachusetts.

BURGESS, J. and J. GOLD (1985): Place, the Media and Popular Culture. J. BURGESS and J. R. GOLD (Eds.): *Geography, the Media and Popular Culture*. London, 1–32.

CALI, E. (2008): Personal communication, February 22, 2008.

CAMPBELL, R. and E. PEREZ (2004): Minorities Dominate 'Real' OC, *Orange County Register*, October 1, 2004. http://www.ocregister.com/ocr/sections/local/local/article_260181.php (accessed August 1, 2008).

CARTER, S. and D. MCCORMACK (2006): Film, Geopolitics and the Affective Logics of Intervention. *Political Geography* 25 (2), 228–245.

CASEY, E. (2002): *Representing Place: Landscape Painting and Maps*. Minneapolis.

CHANG, R. (2004): MTV Loves Laguna: 'The Real Orange County' Will Track Beach City Teens for a Second Season, *Orange County Register*, December 9, 2004. http://infoweb.newsbank.com (accessed December 20, 2007).

CHANG, R. (2007): 'Housewives,' But Not Desperate, *Orange County Register*, January 14, 2007. http://infoweb.newbank.com (accessed December 20, 2007).

CLARKE, D. (1997): Introduction: Previewing the Cinematic City. CLARKE, D. (Ed.): *The Cinematic City*. London, 1–18.

Coldwell Banker (2011): Home Listing Report. http://hlr.coldwellbanker.com/PressRelease.html (accessed September 1, 2011).

CORNER, J. (2002): Performing the Real: Documentary Diversions. *Television and New Media* 3 (3): 255–269.

COSGROVE, D. (1984): *Social Formation and Symbolic Landscapes*. Madison.

COSGROVE, D. (1987): New Directions in Cultural Geography. *Area* 19 (2), 95–101.

COSGROVE, D. (1989): Geography is Everywhere: Culture and Symbolism in Human Landscapes. GREGORY, D. and R. WALFORD (Eds.): *Horizons in Human Geography*. Basingstoke, 118–135.

COSGROVE, D. (1993): Commentary on "The Reinvention of Cultural Geography" by Price and Lewis. *Annals of the Association of American Geographers* 83 (3), 515–516.

COULDRY, N. (1998): The View from Inside the 'Simulacrum': Visitors' Tales from the Set of *Coronation Street*. *Leisure Studies* 17 (2), 94–107.

COULDRY, N. (2005): On the Actual Street. CROUCH, D., R. JACKSON and F. THOMPSON (Eds.): *The Media and the Tourist Imagination: Converging Cultures*. London, 60–75.

CRANG, M., CRANG, P. and J. MAY (1999): Introduction. CRANG, M., CRANG, P. and J. MAY (Eds.): *Virtual Geographies: Bodies, Space and Relations*. London, 1–20.

CRESSWELL, T. (1996): *In Place/Out of Place: Geography, Ideology, and Transgression*. Minneapolis.

CRESSWELL, T. (2003): Landscape and the Obliteration of Practice. ANDERSON, K., D. DOMOSH, S. PILE and N. THRIFT (Eds.): *Handbook of Cultural Geography*. London.

CRESSWELL, T. (2004): *Place: A Short Introduction*. Malden, Massachusetts.

CRESSWELL, T. and D. DIXON. (2002): Introduction: Engaging Film. CRESSWELL, T. and D. DIXON (Eds.): *Engaging Film: Geographies of Mobility and Identity*. Lanham, Maryland, 1–10.

CROUCH, D. and N. LÜBBREN (Eds.) (2003): *Visual Cultural and Tourism*. Oxford.

CURRY, M. (2004): Dallas Tourism Officials Try to Shed City's TV Image, *The Battalion*, April 16, 2004. http://www.thebatt.com/home/indext.cfm (accessed April 7, 2007).

CURTI, G. H., J. DAVENPORT and E. L. JACKIEWICZ (2007): Concrete Babylon: Life Between the Stars: To Dwell and Consume (with)in the Fold(s) of Hollywood, CA. *Association of Pacific Coast Geographers Yearbook* 69, 45–73.

DANIELS, S. (1989): Marxism, Culture, and the Duplicity of Landscape. PEET, R. and N. THRIFT (Eds.): *New Models in Geography*. London, 196–220.

DANIELS, S. and D. COSGROVE (1988): Introduction: Iconography and Landscape. COSGROVE, D. (Ed.): *Iconography of Landscape*. New York, 1–9.

DEMPSEY, J. (2004): MTV Finding Reality in 'Laguna Beach,' *Variety*, June 27, 2004. http://www.variety.com/article/VR117907087.html (accessed January 6, 2008).

DICK, P. K. (1977): *A Scanner Darkly*. New York.

DOEL, M. (2008): From Animated Photography to Film: The Formation of Vernacular Relativity in Early English Films (1895–1908). LUKINBEAL, C. and S. ZIMMERMANN (Eds.): *The Geography of Cinema – A Cinematic World*. Stuttgart, 87–99.

DOEL, M. and D. CLARKE (2007): Afterimages. *Environment and Planning D: Society and Space* 25 (5), 890–910.

DRISCOLL, G. and M. DOSS (2005): Irvine Tops List of Safest U.S. Cities, *Orange County Register*, June 7, 2005. http://www.ocregister.com/ocr/sections/local/local/article_549726.php (accessed September 15, 2008).

DUBOW, J. (2004): Outside of Place and Other Optical: Walter Benjamin and the Geography of Critical Thought. *Journal of Visual Culture* 3 (3), 259–274.

DUNCAN, J, and N. DUNCAN (1988): '(Re)reading the Landscape.' *Environment and Planning D: Society and Space* 6 (2), 117–126.

DUNCAN, J. (1990): *The City as Text: The Politics of Landscape Interpretation in the Kandyan Kingdom*. Cambridge.

DUNCAN, J. (1993): On "The Reinvention of Cultural Geography" by Price and Lewis. *Annals of the Association of American Geographers* 83 (3), 517–519.

DUNCAN, J. and D. LEY (1993): Introduction: Representing the Place of Culture. DUNCAN, J. and D. LEY (Eds.): *Place/Culture/Representation*. London, 1–21.

DUNLOP, S. (2008): Introductory Remarks at Media Alliance of Orange County Luncheon, "Producing Reality TV in Orange County," Costa Mesa, California, February 25, 2008.

ECO, U. (1983): *Travels in Hyperreality*. San Diego.

EISENSTEIN, S. (1987): *NonIndifferent Nature*. Cambridge, Massachusetts.

ESCOBAR, A. (2001): Culture Sits in Places: Reflections on Globalism and Subaltern Strategies of Localization. *Political Geography* 20 (2), 139–174.

FEARY, M. (2008): Personal communication, February 25, 2008.

FEUER, J. (1995): *Seeing Through the Eighties: Television and Reaganism*. Durham.

FISKE, J. (1985): TV and Popular Culture. Paper presented at Iowa Symposium and Conference on Television Criticism, Iowa City, Iowa, April 1985.

FLECKENSTEIN, K. (2006): Between Iconophillia and Iconophobia: Milton's *Areopagitica* and Seventeeth-Century Visual Culture. BIZZELL, P. (Ed.): *Rhetorical Agendas*. Mahwah, New Jersey, 69–86.

FOX (2007): The Sun Sets on *The OC* When the Series Finale Airs Thursday, February 22, on Fox. Fox Press Release, January 3, 2007.

FRIEDMAN, J. (2002): Introduction. FRIEDMAN, J. (Ed.): *Reality Squared: Televisual Discourse on the Real*. New Brunswick, 1–22.

GALLAUGHER, A. (2008): Personal communication, February 22, 2008.

GALLO, P. (2003): The OC, *Variety*, August 3, 2003. http://www.variety.com/index.asp (accessed January 6, 2008).

GILL, A. (2009): The Out-of-Towners, *Vanity Fair*, January 2009, 52–53.

GOLD, J. (1974): Communicating Images of the Environment. *University of Birmingham Centre for Urban and Regional Studies Occasional Papers* 29.

GORDON, D. (2005): Swimming With Sharks: Deconstructing MTV's Teen Phenom 'Laguna Beach,' *Newsweek*, September 26, 2005, 16.

GRIFFIN, S. (2008): Introductory Remarks at Media Alliance of Orange County Luncheon, "Producing Reality TV in Orange County," Costa Mesa, California, February 25, 2008.

HANNA, S. (1996): Is it Roslyn or is it Cicely? Representation and the Ambiguity of Place. *Urban Geography* 17 (7), 633–649.

HARAWAY, D. (1991): *Simians, Cyborgs and Women: The Reinvention of Nature*. New York.

HARTLEY, J. (1996): *Popular Reality*. London.

HARVEY, D. (1990): *The Condition of Postmodernity*. Cambridge, Massachusetts.

HARVEY, S. (2002): Breakers Oust Artists in Laguna Beach; Are Poets and Banana Slugs Next? *Los Angeles Times*, September 11, 2002. http://articles.latimes.com/2002/sep/11/local/me-only11 (accessed November 14, 2010).

HAY, J. (1997): Piecing Together What Remains of the Cinematic City. CLARKE, D. (Ed.): *The Cinematic City*. London, 209–229.

HEATH, S. (1981): *Questions of Cinema*. Bloomington.

HEFFERNAN, V. (2005): Surf's Up! Teenage Hormones are High, *New York Times*, August 12, 2005. http://www.nytimes.com/2005/08/12/arts/television/12heff.html (accessed November 14, 2010).

HIGSON, A. (1984): Space, Place, Spectacle: Landscape and Townscape in the 'Kitchen Sink' Film. *Incorporating Screen Education* 25 (4-5), 2–21.

HIGSON, A. (1987): The Landscape of Television. *Landscape Research* 12 (3), 8–13.

HILL, A. (2002): *Big Brother*: The Real Audience. *Television and New Media* 3 (3): 323–340.

HOLLIER, D. (1998): Premises, Premises: Sketches in Remembrance of a Recent Turn in French Thought. Guggenheim Museum Exhibition Catalogue: *Premises: Invested Spaces in Visual Arts, Architecture and design for France 1958–1998*. New York.

HOLMES, S. and D. JERMYN (2004): Introduction. HOLMES, S. and D. JERMYN (Eds.): *Understanding Reality TV*. New York, 1–32.

HOLTAN, O. (1971): Individualism, Alienation and the Search for Community: Urban Imagery in Recent American Films. *Journal of Popular Culture* 4 (4), 933–942.

HOROVITZ, B. (2002): Color Orange County the New Capital of Cool, *USA Today*, September 5, 2002. http://www.usatoday.com/money/advertising/2002-09-05-orange- County-_x.htm (accessed November 14, 2010).

INGOLD, T. (1993): The Temporality of Landscape. *World Archaeology* 25 (2), 152–174.

ITZKOFF, D. (2008): When Teenage Angst Had its Own ZIP Code, *New York Times*, August 31, 2008. http://www.nytimes.com/2008/08/31/arts/television/31itzk.html (accessed November 14, 2010).

JACKSON, J. (1984): *Discovering the Vernacular Landscape*. New Haven.

JACKSON, P. (1993): Berkeley and Beyond: Broadening the Horizons of Cultural Geography. *Annals of the Association of American Geographers* 83 (3), 519–520.

JACKSON, R. (2005): Converging Cultures; Converging Gazes; Contextualizing Perspectives. CROUCH, D., R. JACKSON and F. THOMPSON (Eds.): *The Media and the Tourist Imagination: Converging Cultures*. London, 183–197.

JANCOVICH, M., L. FAIRE, and S. STUBBINGS (Eds.) (2003): *The Place of the Audience: Cultural Geographies of Film Consumption*. London.

JENCKS, C. (1984): *The Language of Postmodern Architecture*. London.

JHALLY, S. and J. LEWIS (1992): *Enlightened Racism: The Cosby Show, Audiences, and the Myth of the American Dream*. Boulder.

JOYRICH, L. (1996): *Re-Viewing Reception: Television, Gender and Postmodern Culture*. Bloomington.

KELLY, D. (2006): 'Laguna' Cast as the Villain in Laguna, *Los Angeles Times,* November 12, 2006. http://articles.latimes.com/2006/nov/12/local/me-laguna12 (accessed November 14, 2010).

KENNEDY, C. (2008): Living with Film: An Autobiographical Approach. LUKINBEAL, C. and S. ZIMMERMANN (Eds.): *The Geography of Cinema – A Cinematic World.* Stuttgart, 187–204.

KIM, H. and S. L. RICHARDSON. (2003): Motion Picture Impacts on Destination Images. *Annals of Tourism Research* 30 (1), 216–237.

KING, R. (1990): Geographical Insights of the Artist. *Geographical Magazine* 62 (4), 40–42.

KLING, R., S. OLIN and M. POSTER (1991): The Emergence of Postsuburbia: An Introduction. KLING, R., S. OLIN and M. POSTER (Eds.): *Postsuburban California: The Transformation of Orange County since World War II.* Berkeley, 1–30.

KNIGHT, A. (1957): Geography and the Documentary Film: The United States. *The Geographical Magazine* 30 (12), 290–301.

KOMPARE, D. (1994): Extraordinarily Ordinary: The Osbournes as "*An American Family.*" MURRAY, S. and L. OUELLETTE (Eds.): *Reality TV: Remaking Television Culture.* New York, 97–116.

LANSING, D. (2004): Ultimate Orange, *Sunset Magazine*, August 2004. http://www.sunset.com/sunset/travel/article/0,20633,687108,00.html (accessed May 20, 2008).

LEDUFF, C. (2005): Anaheim Journal: Angels Make a Play for Los Angeles in Name Only, *New York Times*, January 6, 2005. http://query.nytimes.com/gst/fullpage.html?res=9905E1D91039F935A35752C0A9639C8B63&pagewanted=all (accessed November 14, 2010).

LEFEBVRE, H. (1991): *The Production of Space*. Translated by Donald Nicholson-Smith. Cambridge, Massachusetts.

LEONCIO, A. (2008): Personal communication, February 21, 2008.

LEVINE, E. (2006): The New Soaps? *Laguna Beach, The Hills*, and the Gendered Politics of Reality "Drama," *FlowTV*, August 18, 2006. http://flowtv.org/?p=19 (accessed November 14, 2010).

LEWIS P. (1979): Axioms for Reading the Landscape: Some Guides to the American Scene. MEINIG, D. (Ed.): *The Interpretation of Ordinary Landscapes: Geographical Essays*. New York, 11–32.

LEWIS, J. (2004): The Meaning of Real Life. MURRAY, S. and L. OUELLETTE (Eds.): *Reality TV: Remaking Television Culture.* New York, 288–302.

LIMSKY, D. (2008): California, Here We Come: The Real 'OC.' *National Geographic Traveler*, October 2008, 147–148.

LOWE, P. (2008): Board Approves 'OC Brand' for Agencies, *Orange County Register*, March 18, 2008. http://www.ocregister.com/articles/supervisors-brand-County-2001505-agencies-government (accessed November 14, 2010).

LUKINBEAL, C. (2004): The Rise of Regional Film Production Centers in North America, 1984–1997. *Geojournal* 59, 307–321.

LUKINBEAL, C. (2005): Cinematic Landscapes. *Journal of Cultural Geography* 23 (1), 3–22.

LUKINBEAL, C. (2006): Runaway Hollywood: Cold Mountain Romania. *Erdkunde* 60 (4), 337–345.

LUKINBEAL, C. and S. ZIMMERMANN (2006): Film geography: A new subfield. *Erdkunde* 60 (4), 315–326.

LUKINBEAL, C. and S. ZIMMERMANN (2008): A Cinematic World. LUKINBEAL, C. and S. ZIMMERMANN (Eds.): *The Geography of Cinema – A Cinematic World.* Stuttgart, 15–23.

MACCANNELL, D. (1992): *Empty Meeting Grounds.* New York.

MACIONIS, N. (2004): Understanding the Film-Induced Tourist. FROST, W., G. CROY and S. BEETON (Eds.): *International Tourism and Media Conference Procdeedings, November 2004.* Melbourne, 86–97.

MALLORY, W. and P. SIMPSON-HOUSLEY (Eds.) (1987): *Geography and Literature: A Meeting of the Disciplines.* Syracuse.

MALRAUX, A. (1967): *Museum without Walls*. Translated by Stuart Gilbert and Francis Price. New York.

MANVELL, R. (1956): Geography and the Documentary Film. *The Geographical Magazine* 29 (9), 417–422.

MARTIN, J.-C. (2000): Of Images and Worlds. FLAXMAN, G. (Ed.): *The Brain is the Screen: Deleuze and the Philosophy of Cinema*. Minneapolis, 61–85.

MASSEY, D. (1994): *Space, Place and Gender*. Minneapolis.

MASSEY, D. (2005): *For Space*. London.

MASSUMI, B. (2002): *Parables for the Virtual: Movement, Affect, Sensation*. Durham, North Carolina.

MATLESS, D. (1992): An Occasion for Geography: Landscape, Representation, and Foucault's Corpus. *Environment and Planning D: Society and Space* 10 (1), 41–56.

MCARTHUR, C. (1997): Chinese Boxes and Russian Dolls: Tracking the Elusive Cinematic City. CLARKE, D. (Ed.): *The Cinematic City*. London, 19–45.

MCCARTHY, A. (2004): Laguna Beach, *FlowTV*, December 3, 2004. http://flowtv.org/?p=667 (accessed November 14, 2010).

MCGIRR, L. (2001): *Suburban Warriors: The Origins of the New American Right*. Princeton.

MCHUGH, K. E. (2009): Movement, Memory, Landscape: An Excursion in Non-Representational Thought. *GeoJournal* 74, 209–218.

MCLAUGHLIN, H. (1975): The Disquieting Roles that Architecture Plays in the Movies. *American Institute of Architects Journal* 63 (1), 39–41.

MEINIG, D. (Ed.) (1979): *The Interpretation of Ordinary Landscapes: Geographical Essays*. New York.

MICKADEIT, F. (2006a): It Takes All Sorts to Make Coto What It Is, *Orange County Register*, April 4, 2006. http://infoweb.newsbank.com (accessed December 20, 2007).

MICKADEIT, F. (2006b): 'Real' Awards: May I Have the Envelope, Please? *Orange County Register*, May 3, 2006. http://infoweb.newsbank.com (accessed December 20, 2007).

MICKADEIT, F. (2006c): Bravo's Perception Violates My Reality, *Orange County Register*, March 31, 2006. http://infoweb.newsbank.com (accessed December 20, 2007).

MITCHELL, D. (2000): *Cultural Geography: A Critical Introduction*. Oxford.

MITCHELL, D. (2008): New Axioms for Reading the Landscape: Paying Attention to Political Economy and Social Justice. WESCOAT, Jr., J. L. and D. M. JOHNSTON (Eds.): *Political Economies of Landscape Change*. London, 29–50.

MITCHELL, W. (1994): Imperial Landscape. MITCHELL, W. J. T. (Ed.): *Landscape and Power*. Chicago, 5–34.

MOORE, G. and R. GOLLEDGE (Eds.) (1976): *Environmental Knowing*. Stroudsburg, Pennsylvania.

MORLEY, D. (1989): Changing Paradigms in Audience Studies. SEITER, E., H. BORCHERS, G. KREUTZNER and E.-M. WARTH (Eds.): *Remote Control: Television, Audiences, and Cultural Power*. London, 16–43.

MORRISON, P. (1990): L.A. vs. O.C.: Just How Different Are They, Anyway? Now the Truth Can Be Told, *Los Angeles Times Magazine*, June 17, 1990, 10–18.

MULVEY, L. (1975): Visual Pleasure and Narrative Cinema. *Screen* 16 (3), 6–18.

NATTER. W. and J.P. JONES, III. (1993): Pets or Meat: Class, Ideology, and Space in *Roger & Me*. *Antipode* 25 (2), 140–158.

Nielsen Media Research (2007): Nielsen Reports Television Tuning Remains at Record Levels, October 17, 2007. http://www.nielsen.com/media/2007/pr_071017a.html (accessed November 21, 2008).

OUELLETTE, L. and S. MURRAY (2004): Introduction. MURRAY, S. and L. OUELLETTE (Eds.): *Reality TV: Remaking Television Culture*. New York, 1–15.

OVERLEY, J. (2007): Logo, Slogan in the Works for O.C., *Orange County Register*, December 6, 2007. http://www.ocregister.com/money/County-orange-effort-1936821-marketing-name (accessed November 14, 2010).

OVERLEY, J. (2008): Are We On TV Too Much? *Orange County Register*, January 4, 2008. http://www.ocregister.com/article/County-orange-dean-1954473-little-among (accessed November 14, 2010).

OVERLEY, J. and R. CHANG (2007): MTV in Talks with Newport, *Orange County Register*, March 28, 2007. http://infoweb.newsbank.com (accessed February 26, 2008).

PARKER, T. J. (1985): *Laguna Heat.* New York.

PARKER, T. J. (1991): *Pacific Beat.* New York.

PECKMAN, R. (2004): Landscape in Film. DUNCAN, J., N. JOHNSON and R. SCHEIN (Eds.): *A Companion to Cultural Geography*. London, 420–429.

PEYSER, M. (2005): Changing Their Tune, *Newsweek*, October 17, 2005, 16.

POCOCK, D. (1981a): *Humanistic Geography and Literature: Essays on the Experience of Place.* London.

POCOCK, D. (1981b): Imaginative Literature and the Geographer. POCOCK, D. (Ed.): *Humanistic Geography and Literature: Essays on the Experience of Place.* London, 9–19.

POORE, B. and N. CHRISMAN (2006): Order from Noise: Towards a Social Theory of Geographic Information. *Annals of the Association of American Geographers* 96 (3), 508–523.

POSTER, M. (1988): Introduction. POSTER, M. (Ed.): *Jean Baudrillard: Selected Writings*. Stanford, California, 1–9.

PRESS, A. (1991): *Women Watching Television*. Philadelphia.

PRICE M. and M. LEWIS. (1993a): The Reinvention of Cultural Geography. *Annals of the Association of American Geographers* 83 (1), 1–17.

PRICE M. and M. LEWIS (1993b): Reply: On Reading Cultural Geography. *Annals of the Association of American Geographers* 83 (3), 520–522.

REES, R. (1973): Geography and Landscape Painting: An Introduction to a Neglected Field. *Scottish Geographical Magazine* 89 (3), 147–157.

REID, S. M. (2008): Why are we home to so many Olympians? *Orange County Register*, July 5, 2008. http://www.ocregister.com/articles/olympic-County-orange-2085530-gold-viejo (accessed November 14, 2010).

RELPH, E. (1976): *Place and Placelessness*. London.

RILEY, R. and C. VAN DOREN (1992): Movies as Tourism Promotion: A 'Pull' Factor in a 'Push' Location. *Tourism Management* 13 (3), September 1992, 267–274.

RILEY, R., D. BAKER and C. VAN DOREN (1998): Movie Induced Tourism. *Annals of Tourism Research* 25 (4), 919–935.

RITCHIE, E. and J. ZLOTNIK. (2004): Laguna Residents Get Real, *Orange County Register*, September 30, 2004. http://infoweb.newsbank.com (accessed December 20, 2007).

ROBERTSON, I. and P. RICHARDS (2003): Introduction to Studying Cultural Landscapes. ROBERTSON, I. and P. RICHARDS (Eds.): *Studying Cultural Landscapes*. London, 1–18.

ROBINSON, K. S. (1988): *The Gold Coast.* New York.

ROCHLIN, M. (2005): An MTV Coming of Age That Went Far on Charm, *New York Times*, August 30, 2005. http://www.nytimes.com/2005/08/30/arts/television/30lagu.html (accessed November 14, 2010).

ROJEK, C. (1997): Indexing, Dragging, and the Social Construction of Tourist Sights. ROJEK, C. and J. URRY (Eds.): *Touring Cultures: Transformations of Travel and Theor.* New York, 52–74.

ROSE, G. (1993): *Feminism and Geography: The Limits of Geographical Knowledge.* Minneapolis.

ROSE, G. (2001): *Visual Methodologies: An Introduction to the Interpretation of Visual Materials.* London.

ROSE, M. (2002): Landscape and Labyrinths. *Geoforum* 33 (4), 455–467.

ROSE, M. (2006): Gathering 'Dreams of Presence': A Project for the Cultural Landscape. *Environment and Planning D: Society and Space* 24 (4), 537–554.

ROSENTHAL, E. (2003): North of Beijing, California Dreams Come True, *New York Times*, February 3, 2003. http://query.nytimes.com/gst/fullpage.html (accessed November 14, 2010).

SADLER, W. J. and E.V. HASKINS (2005): Metonymy and the Metropolis: Television Show Settings and the Image of New York City. *Journal of Communication Inquiry* 29 (3), 195–216.

SALTER, C. and W. LLOYD (1976): *Landscape and Literature Resource Paper No. 76–3*. Washington D.C.

Santa Margarita Company (1986): *Rancho Santa Margarita Community Brochure.*

SCHEIBAL, M. (2008): Introductory Remarks at Media Alliance of Orange County Luncheon, "Producing Reality TV in Orange County," Costa Mesa, California, February 25, 2008.

SCHOENKOPF, R. (2008): *Commie Girl in the OC.* London.

SCHWARTZ, J. (2004): *Charlie Rose.* Interview by Charlie Rose. November 22, 2004. http://www.charlierose.com/shows/2004/11/22/2/a-conversation-with-the-o-c-creator-josh-schwartz (accessed May 14, 2008).

SCOTT, A. (1988): *Metropolis: From Division of Labor to Urban Form.* Berkeley.

SEIGEL, A. (2004): *Like the Red Panda.* New York.

SHAVIRO, S. (1993): *The Cinematic Body.* Minneapolis.

SHAW, I. and B. WARF (2009): Worlds of Affect: Virtual Geographies of Video Games. *Environment and Planning A* 41 (6), 1332–1343.

SHERMAN, S. (1967): How the Movies See the City. *Landscape* 16 (3), 25–26.

SHIEL, M. and T. FITZMAURICE (Eds.) (2001): *Cinema and the City: Film and Urban Studies in a Global Context.* Oxford.

SHOUSE, E. (2005): Feeling, Emotion, Affect. *M/C Journal* 8 (6). http://journal.media-culture.org.au/0512/03-shouse.php (accessed November 14, 2010).

SICKLER, S. (2007a): Behind the Gates: Coto Gains New Popularity, *Orange County Register*, December 12, 2007. http://www.ocregister.com/articles/housewives-real-columns-1940507-orange-County (accessed November 14, 2010).

SICKLER, S. (2007b): Behind the Gates: What You Don't See on Reality TV. *Orange County Register*, 26 December. http://www.ocregister.com/news/jeana-174381-people-don.html_(accessed 14 November 2010).

SICKLER, S. (2008): Behind the Gates: Dispelling Myths About Coto, *Orange County Register*, January 19, 2008. http://www.ocregister.com/articles/really-coto-gated-1962712-homes-around (accessed November 14, 2010).

SINGH, S. (2004): *Goddess for Hire.* New York.

SMITH, L. (2005): There's Laguna, and Then There's MTV's 'Laguna," *Los Angeles Times*, November 5, 2005. http://pqasb.archiver.com (accessed January 18, 2008).

SMITH, L. (2006): Residents See Red Over Orange County 'Housewives.' *Los Angeles Times,* April 3, 2006. http://infoweb.newsbank.com (accessed December 20, 2007).

SMITH, L. (2007): Busting Out of The O.C., *Los Angeles Times*, March 16, 2007. http://infoweb.newsbank.com (accessed December 20, 2007).

SMITH, N. and C. KATZ (1993): Grounding Metaphor: Towards a Spatialized Politics. KEITH, M. and S. PILE (Eds.): *Place and the Politics of Identity.* New York, 67–83.

SOJA, E. W. (1996): *Thirdspace: Journeys to Los Angeles and Other Real-and-Imagined Places.* Malden, Massachusetts.

SPIGEL, L. (2002): The Suburban Home Companion: Television and the Neighborhood Ideal in Postwar America. JONES, A. (Ed.): *The Feminism and Visual Culture Reader (Sight: Visual Culture)*, New York, 329–336.

STEINHAUER, J. (2006): Real 'OC' Starts Objecting to Its MTV Portrayal, *New York Times*, November 8, 2006. http://www.nytimes.com/2006/11/08/us/08laguna.html (accessed November 14, 2010).

STENGER, J. (2001): Return to Oz: The Hollywood Redevelopment Project, or Film History as Urban Renewal. SHIEL, M. and T. FITZMAURICE (Eds.): *Cinema and the City: Film and Urban Studies in a Global Context.* Oxford, 59–72.

STRAIGHT, S. (2003): Dissed by 'The O.C.,' *Salon.com*, September 9, 2003. http://dir.salon.come /story/mwt/feature/2003/09/09/oc/index.html (accessed September 30, 2008).

STREISAND, B. (2003): Hip's New Headquarters, *US News and World Report*, August 20, 2003. http://www.usnews.com/usnews/news/articles/030818/18orange_print.htm (accessed November 14, 2010).

THRIFT, N. (1999): Steps to an Ecology of Place. MASSEY, D., J. ALLEN and P. SARRE (Eds.): *Human Geography Today.* Cambridge, 295–322.

THRIFT, N. (2008): *Non-representational Theory: Space, Politics, Affect.* London.

TILL, K. (1993): Neotraditional Towns and Urban Villages: The Cultural Production of a Geography of 'Otherness.' *Environment and Planning D: Society and Space* 11 (6), 709–732.

TOOKE, N. and M. BAKER (1996): Seeing is Believing: The Effect of Film on Visitor Numbers to Screened Locations. *Tourism Management* 17 (2), 87–94.

TORCHIN, L. (2002): Location, Location, Location: The Destination of the Manhattan TV Tour. *Tourist Studies* 2 (3), 247–266.

TRANI, B. (2008): Personal communication, December 10, 2008.

TUAN, Y. (1990): Realism and Fantasy in Art, History, and Geography. *Annals of the Association of American Geographers* 80 (3), 435–446.

TUAN, Y. (1977): *Space and Place: The Perspective of Experience*. Minneapolis.

TUAN, Y. (1991): Language and the Making of Place: A Narrative-Descriptive Approach. *Annals of the Association of American Geographers* 81 (4), 684–696.

U.S. Census Bureau (2000): American Factfinder. http://factfinder.census.gov (accessed September 1, 2011).

VENTAKESH, A. (1991): Changing Consumption Patterns. KLING, R., S. OLIN and M. POSTER (Eds.): *Postsuburban California: The Transformation of Orange County since World War II.* Berkeley, 142–164.

VH1 (2002): Agents Orange! First Hand Reports from Orange County. http://www.vh1.com/ shows/dyn/vh1_news_special/57487/episode_interviews_int.jhtml?start=1 (accessed September 14, 2008).

VILLARREAL, Y. and A. POWERS (2006): Sunset for 'The O.C.'? That Depends, *Los Angeles Times*, December 12, 2006. http://www.latimes.com (accessed April 2, 2007).

VLAHIDES, J. A. (2007): *Coastal California*. Melbourne.

WEIR, K. (2004): *Moon Handbooks Coastal California*. Berkeley, California.

WILLIAMS, A. (2004): The County Formerly Known as Squaresville, *New York Times*, October 10, 2004. http://www.nytimes.com/2004/10/10/fashion/10ORAN.html (accessed January 2, 2008).

WIRTH, E. (1952): *Stoffprobleme des Films*. Inaugural-Dissertation. Freiburg.

WYATT, E. (2007): 'The OC': A Fast Start, A Faster Finish, *New York Times*, January 6, 2007. http://www.nytimes.com/2007/01/06/arts/television/06josh.html (accessed November 14, 2010).

WYLIE, J. (2006): Depths and Folds: On Landscapes and the Gazing Subject. *Environment and Planning D: Society and Space* 24 (4), 519–535.

YOSHINO, K. (2003): Suddenly, O.C. Leads the Hip Parade, *Los Angeles Times*, July 19, 2003 http://articles.latimes.com/2003/jul/19/local/me-hipoc19 (accessed November 14, 2010).

ZONN, L. (1990): *Place Images in Media: Portrayal, Experience and Meaning.* Savage, Maryland.

APPENDIX A

MIDWESTERN STUDENT SURVEY INSTRUMENT

1. Are you male or female?

2. Where are you from?

3. Please list 5 words you would use to describe Orange County, California.

    a)
    b)
    c)
    d)
    e)

4. Have you ever traveled to Orange County, California?

    a) If yes, please describe your trip. What was the purpose? What did you see? What activities did you do?
    b) How would you describe Orange County? Is it a desirable place to visit? What does it look like? What are the people like? What is the lifestyle like?

5. If you haven't been to Orange County, have you ever traveled to Southern California? If so, where?

6. How would you characterize your travel habits?

    a) I travel outside of the state several times per year.
    b) I travel outside of the state a few times per year.
    c) I rarely leave the state.
    d) I rarely leave the immediate area in which I live.

    Please elaborate on your answer to Question 6. What kinds of trips do you take? Where do you generally go?

7. How would you characterize your TV viewing habits?

a) I watch TV more than three hours per day.
b) I watch TV between 1 and 3 hours per day.
c) I watch TV less than 1 hour per day.
d) I rarely watch TV.

8. If you watch TV, what types of shows do you enjoy watching the most? Please select your top 2.

    a) Hour-long melodramas (e.g. *Grey's Anatomy*)
    b) Procedural crime dramas (e.g. *CSI*, *Law & Order*)
    c) Comedies (e.g. *The Office*)
    d) Documentary-style reality shows (i.e. cameras follow people around) (e.g. *The Real World*, *The Hills*)
    e) Competition-style reality shows (e.g. *Survivor*, *American Idol*, *The Bachelor*)
    f) Talk shows (e.g. *Regis & Kelly*, *Oprah*)
    g) Game shows (e.g. *Deal or No Deal*)
    h) Daytime soaps
    i) News
    j) Other, please describe ____________________.

9. Please list the current serial shows you follow regularly, if any.

10. Do you/did you watch the following TV shows:

    a) *The OC*
        i. Regularly
        ii. I've seen it a few times
        iii. Never
    b) *Laguna Beach: The Real Orange County*
        i. Regularly
        ii. I've seen it a few times
        iii. Never
    c) *The Real Housewives of Orange County*
        i. Regularly
        ii. I've seen it a few times
        iii. Never
    d) Have you seen any other movies or TV shows set in Orange County? If so, please list.

11. If you haven't watched any of these shows, are you aware of them? Are you generally familiar with their subject matter?

a) If so, please describe what you know of these shows and how you know it.
b) From your knowledge of these shows, do you think they are representative of Orange County?

12. Would you consider yourself a fan of any one of these shows? If so, which one(s)?

a) If you consider yourself a fan, what do you like most about the show(s)?
b) If you consider yourself a fan, can you identify any effects the show(s) have had on you? For example, have they influenced the music you listen to, the clothes you wear, or your speech?
c) Do you/did you get together to watch any of these shows with a group? If so, which one(s)?

13. In your opinion, how does the setting (Orange County) contribute to the appeal of the show(s)? What qualities, if any, does the setting provide to make the show more enjoyable or entertaining?

14. Which most accurately characterizes your prior knowledge of Orange County, California, before seeing these shows? Please choose all that apply.

a) I had visited Orange County before.
b) Although I haven't been there, I am familiar with it from another source (e.g. books, music)
c) I know where Orange County is located.
d) I had heard of Orange County, but didn't know much about it.
e) I had not heard of Orange County before seeing these shows.

Please elaborate on your answer to Question 14.

15. How much of your knowledge about Orange County comes from these shows?

a) Very much to all. I was not familiar with Orange County before I watched these shows.
b) Some. I knew something of Orange County before I watched these shows.
c) Very little to none. I was very familiar with Orange County before I watched these shows.

16. On a scale of 1–5 (1 being the least), how much of your impression of Orange County would you say has been formed by these shows?

1 2 3 4 5

Please elaborate on your answers to Questions 15 & 16.

17. Have these shows changed your impression of Orange County?

    a) No. I didn't have an impression of Orange County before I saw them.
    b) No. My impression of Orange County didn't change after I saw them.
    c) Somewhat.
    d) Yes. These shows formed my impression of Orange County.

    Please elaborate on your answer to Question 17. What is your current impression of Orange County?

18. How would you describe the image of Orange County that these shows project?

19. How would you describe the lifestyle of the characters on these shows, or the 'Orange County' lifestyle?

20. How accurately do you think these shows portray Orange County?

    a) Very accurately.
    b) Somewhat accurately, especially the reality shows.
    c) Somewhat inaccurately.
    d) Very inaccurately.
    e) I couldn't say.

    Please elaborate on your answer to Question 20.

21. Do you think these shows portray Orange County in a realistic manner? Why or why not?

22. Is Orange County a place similar to or different from where you live? In what ways? What are the positive and negative characteristics of Orange County as seen on these shows?

23. Are the lives depicted on the shows similar to or different from yours? In what ways? Would you like to live the 'Orange County' lifestyle?

24. Based on what you've seen in these shows, is Orange County a place you would like to visit, or even to live? Why or why not?

25. Have you actually visited Orange County to see the place portrayed on these shows? If so, what motivated you to make the trip? How did you feel the actual place compared to what you've seen on TV?

APPENDIX B

FAN SURVEY INSTRUMENT

1. Are you male or female?

2. What year were you born?

3. Where are you from?

4. Please list 5 words you would use to describe Orange County, California.

5. Please list 5 words you would use to describe the place where you live.

6. Which of the following TV show(s) did you watch regularly?

    a) *The OC*
    b) *Laguna Beach: The Real Orange County*
    c) *The Real Housewives of Orange County*

7. Of which of the following TV shows would you call yourself a fan?

    a) *The OC*
    b) *Laguna Beach: The Real Orange County*
    c) *The Real Housewives of Orange County*

8. What do you like most about the show?

9. Do you feel that the setting (Orange County) was important to your enjoyment of the show(s)?

10. If yes, why?

11. Describe how Orange County was portrayed on the show(s).

12. Is Orange County a place you find appealing?

13. How accurately do you think The OC portrays Orange County?

a) Very accurately.
b) Somewhat accurately, especially the reality shows.
c) Somewhat inaccurately.
d) Very inaccurately.
e) I couldn't say.

Please elaborate on your answer to Question 13.

14. Which best characterizes your prior knowledge of Orange County, California, before seeing these shows?

a) I had been to Orange County before.
b) Although I haven't been there, I am familiar with it from another source (e.g. books, music).
c) I knew where Orange County is located.
d) I had heard of Orange County, but didn't know much about it.
e) I had not heard of Orange County before seeing these shows.

15. How much of your knowledge about Orange County comes from these shows?

a) Very much to all. I was not familiar with Orange County before I watched these shows.
b) Some. I knew something of Orange County before I watched these shows.
c) Very little to none. I was very familiar with Orange County before I watched these shows.

16. Have these shows changed your impression of Orange County?

a) No. I didn't have an impression of Orange County before I saw them.
b) No. My impression of Orange County didn't change after I saw them.
c) Somewhat.
d) Yes. These shows formed my impression of Orange County.

Please elaborate on your answers to Questions 14–16.

17. Have you ever traveled to Orange County?

18. If yes, did you travel to Orange County before or after seeing these shows?

19. How did the actual Orange County compare to what you've seen on TV?

20. Was your visit a direct result of seeing Orange County on television?

a) Yes
b) No
c) Partly

21. If yes or partly, what TV-related sites did you see? Were you satisfied with your Orange County experience? Why or why not?

APPENDIX C

ORANGE COUNTY STUDENT SURVEY INSTRUMENT

1. Are you male or female?

2. In which city do you live?

3. Please list 5 words you would use to describe Orange County.

   a)
   b)
   c)
   d)
   e)

4. Do you/did you watch the following TV shows:

   a) *The OC*
      iv. Regularly
      v. I've seen it a few times
      vi. Never
   b) *Laguna Beach: The Real Orange County*
      iv. Regularly
      v. I've seen it a few times
      vi. Never
   c) *The Real Housewives of Orange County*
      iv. Regularly
      v. I've seen it a few times
      vi. Never

5. Do you consider yourself a fan of any of these shows? If so, which one(s)?

6. How accurately do these shows portray Orange County?

   a) Very accurately.
   b) Somewhat accurately.
   c) Somewhat inaccurately.

d) Very inaccurately.
e) I couldn't say.

7. Please elaborate on your answer to Question 6. Do you think these shows portray Orange County, or the particular community they are set in (i.e. Newport, Laguna, or Coto de Caza), in a realistic manner? Why or why not?

8. Do you think the reality shows *Laguna Beach* or *The Real Housewives* portray Orange County in a more realistic manner than *The OC*? Why or why not?

9. How would you describe the image of Orange County that these shows project?

10. In your opinion, how does the Orange County setting contribute to the appeal of the shows? What qualities does the setting provide to make the show more enjoyable or entertaining?

11. Is the Orange County of television a place similar to or different from where you live? In what ways?